Iontronics

Ionic Carriers in Organic Electronic Materials and Devices

Iontronics

Ionic Carriers in Organic Electronic Materials and Devices

Edited by

Janelle Leger
Magnus Berggren
Sue Carter

CRC Press
Taylor & Francis Group
Boca Raton   London   New York

CRC Press is an imprint of the
Taylor & Francis Group, an **informa** business

CRC Press
Taylor & Francis Group
6000 Broken Sound Parkway NW, Suite 300
Boca Raton, FL 33487-2742

First issued in paperback 2017

© 2011 by Taylor and Francis Group, LLC
CRC Press is an imprint of Taylor & Francis Group, an Informa business

No claim to original U.S. Government works

ISBN 13: 978-1-138-11650-4 (pbk)
ISBN 13: 978-1-4398-0688-3 (hbk)

Library of Congress Cataloging-in-Publication Data

Iontronics : Ionic carriers in organic electronic materials and devices / edited by Janelle Leger, Magnus Berggren, Sue Carter."
    p. cm.
  Includes bibliographical references and index.
  ISBN 978-1-4398-0688-3 (hardcover : alk. paper)
  1. Organic electronics. 2. Ionophores. I. Leger, Janelle. II. Berggren, Magnus. III. Carter, Sue A., 1966- IV. Title.

TK7871.I69 2011
621.382--dc22
                                                  2010030233

Visit the Taylor & Francis Web site at
http://www.taylorandfrancis.com

and the CRC Press Web site at
http://www.crcpress.com

Contents

Foreword

Qibing Pei

Iontronics deals with the phenomena and devices that involve the flow of both electrons and ions. Organic compounds and polymers consisting of conjugated electronic structures can conduct electrons. Charging or discharging these materials entails ionic carriers moving in or out of the materials. The electrochemical process can cause a number of important properties to change, including electronic conductivity, color, fluorescence intensity, and volume. Therefore, one may use electricity conveniently to modulate a specific material property. Studying the underlying ionic flow has lead to new scientific understandings and exciting application devices. Publications on this subject have been scattered in the literature due to the multidisciplinary nature of the research and the researchers as well as the diversity of the applications. This book, devoted exclusively to the iontronics in organic electronic materials, therefore seems timely. The Foreword is a good place to take a big picture look at where the field has been and where it is going.

NON-CONJUGATED IONTRONIC MATERIALS

Iontronics could be dated back to as early as the discovery of electricity in the 18th century. The work by Luigi Galvani in the late 1790s on the connection between muscle contraction and electricity marked the birth of electrochemistry. Today, the electrochemical process in neuromuscular systems remains an important research topic. A number of electrochemical devices, such as batteries and electroplated metallic coatings, have been indispensible parts of modern life. Most of the devices are based on metals or ceramic crystals that are electrically conductive. When placed in electrolyte solutions, electrochemical reduction and oxidation can take place under the influence of an electric field. Ion diffusion often plays a controlling role in the reaction rate. Organic materials, on the other hand, had been considered as electrically inactive as they generally lack free electrons or cannot conduct electrons over a long range. The electrochemistry of organic compounds dissolved in liquid electrolyte solutions, however, has a long history, mostly for the electrosynthesis of new molecules. There were also reports of observations of interesting photophysical phenomena. A classic example is the electrogenerated chemiluminescence in organic solutions that was pioneered by Hercules,[1] Chandross and Sonntag,[2] and Santhanam and Bard.[3] The phenomenon was suggested for developing passive matrix displays, although the technological effort has abated since the early 1990s due to the difficulty in encapsulating volatile solvents and the emergence of active matrix liquid crystal displays.

The electrogenerated chemiluminescence in a thin solid film of poly[tris(4-vinyl-4'-methyl-2,2'-bipyridyl)ruthenium(II)] merits further notation.[4] The polymer film was used as a modified electrode to achieve light emission in an acetonitrile solution.

More recently, solid-state electroluminescent devices have been reported using transition metal complexes as the emissive material. In a series of publications, Rubner et al. showed that a thin film of tris(2,2'-bipyridyl) ruthenium(II) complex dispersed in a polymer matrix and sandwiched between indium-tin oxide and metallic electrodes could yield fairly high photon/electron quantum efficiency and modest brightness at low driving voltages.[5] The device performance has been further increased using different ionic carriers or transitional metal complexes. Electronic conduction via the organometallic materials is essential for luminescence to sustain, although how electrons are conducted remains to be elucidated. Involvement of ionic carriers was suggested to be a key in the electroluminescence.

EVENTS PRIOR TO CONDUCTING POLYACETYLENE

The passage of electric current through an organic material was first reported by Kallmann and Pope in 1960.[6] They observed large changes in bulk conductivity in an anthracene crystal in contact with iodine. The conductivity increase was explained by electron extraction from the crystal "injecting a positive hole inside the anthracene which would travel under the action of the field through the crystal to the negative electrode (p. 301)." This seminal work encouraged the use of organic materials, which had been regarded as non-conductive, for electronic devices. Pope et al. further observed electroluminescence from the crystal in 1963.[7] They used silver paste or 0.1M NaCl aqueous solutions as the charge injection electrodes. At 400 V applied across the crystal about 10 to 20 μm thick, light emission was generated that resembled the crystal's fluorescence. In single crystal anthracene with 0.1 mol% tetracene, the luminescence was the fluorescence of the tetracene impurity. This was done 24 years before Tang and VanSlyke[8] reported high efficiency organic thin film electroluminescent diodes that sparked the commercial exploration of organic light emitting diodes (OLEDs).

Another important chapter in the history of organic conductors was organic salts based on TCNQ (7,7,8,8-tetracyano-p-quinodimethane) and TTF (tetrathiofulvalene).[9,10] Electron transfer between this acceptor-donor pair resulted in the formation of a metallic state in the crystal. The crystal is as electrically conducting as a metal, σ (298 K) = 6 × 10^5 ohm^{-1} cm^{-1}. Superconductivity was obtained in (TMTSF)$_2$PF$_6$ (di-(tetramethyltetraselenafulvalene)-hexafluorophosphate), known as Bechgaard salt.[11] The research on organic charge transfer salt and un-charged crystals had been active until around the end of the 1970s when the discovery of high conductivity polyacetylene caught a broad interest in conducting polymers. Organic (semi)conductors regained their luster after the Tang and VanSlyke paper was published.

Before polyacetylene became the characteristic conducting polymer, several other conducting polymers had already been reported. The best known was polysulfurnitride, (SN)$_x$, a one-dimensional inorganic polymer chain with metallic conductivity in a broad temperature range of 4.2 to 300 K.[12] (SN)$_x$ is a neutral polymer. Charging or doping is not required to obtain the high electronic conductivity. Nowak et al. first used (SN)$_x$ as the electrode for the cyclic voltammetry of Pb(NO$_3$)$_2$.[13] It was found that this polymeric electrode material was unique in its strong surface interactions with ions in the electrolyte phase. They suggested that the polymer surface might be more amenable to chemical modification for the production of redox catalysts than graphite or noble metal electrodes. This concept was later realized on conjugated polymers.

ELECTRON CONJUGATED POLYMERS

Polyacetylene has alternating carbon-carbon single and double bonds, each carbon carries a π-electron that shares with all other π-electrons on the same chain. This conjugated structure was predicted by Little as an ideal organic conductor.[14] Heeger, MacDiarmid, and Shirakawa et al. were accredited with the discovery of conducting polymers for doping polyacetylene films with bromine or iodine to increase the polymer's conductivity by over 9 orders of magnitude.[15] This 1977 discovery immediately caused wide-spread interest in the search for stable and highly conductive polymers containing a π-electron conjugated main chain. A number of conjugated polymers were discovered or re-discovered (Figure 1). These organic polymers all show metallic conductivity when heavily doped. Solid-state physics and experimental analysis seem to point to charged solitons being the charge carriers in poly(trans-acetylene), which has a degenerate ground state. The rest of the conjugated polymers shown in Figure 1 have a non-degenerate ground state. The consensus is that polarons are the dominant charge carriers in the lightly doped polymers, and bipolarons are responsible for conducting electricity in heavily doped polymers. The conductivity, although metallically high, increases with temperature, characteristic of a hopping mechanism predicted by the Anderson-type variable range hopping in amorphous solids.

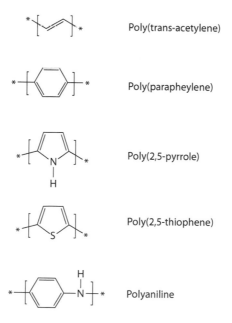

Poly(trans-acetylene)

Poly(parapheylene)

Poly(2,5-pyrrole)

Poly(2,5-thiophene)

Polyaniline

FIGURE 1   Representative iontronic polymers.

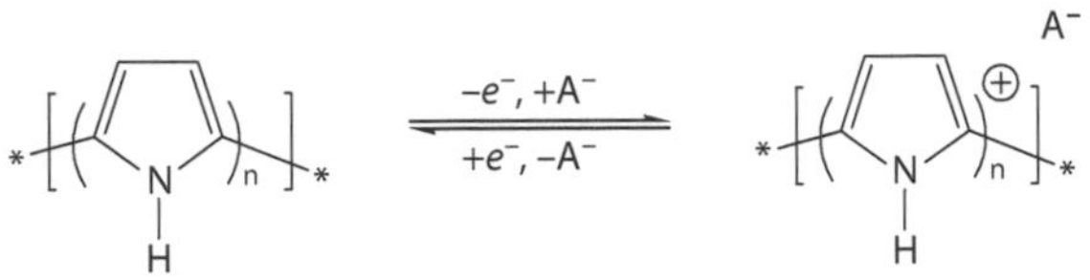

FIGURE 2 Illustration of electrochemical switching between neutral and heavily doped polypyrrole.

The doping in the conjugated polymers is different in the sense of inorganic semi-conductors. The doped polymers are densely charged, with the charges being compensated by counter-ions. The structure shown on the right hand of the equation in Figure 2 is for a heavily p-doped polypyrrole. Roughly every three to four pyrrole units (n = 3 or 4) carry a positive charge, which is compensated by an anion (A⁻). Diaz and Castillo were the first to realize that polypyrrole can be electrochemically switched between the neutral, insulator state and the doped, conductive state (Figure 2).[16] The switching is reversible for numerous cycles. The polymer electrode was stable and non-porous, and it was possible to obtain voltammograms with it. It has been found by now that almost all conjugated polymers have this property. Electrochemistry therefore becomes a convenient tool to introduce a variety of anionic dopants. It has also been used extensively to characterize these materials, including their ionization potential and electron affinity.

Nigrey et al. first showed in 1981 that the reversible switching of polyactylene could be useful for energy storage.[17] The energy stored in a doped polyacetylene was 176 W-h/kg, which compared well with the figure-of-merit of 30 W-h/kg for lead acid batteries used in automobles. Polymer rechargeable batteries attracted great interest in the 1980s. However, this sub-field has subdued since the 1990s due to limited cycle lifetime and energy density compared to the emerging lithium polymer batteries. The understanding gained on the charging, discharging, and ionic transport in the conjugated polymers has benefited the study of organic electronic devices.

ELECTROCHROMISM

The electrochemical switching between the neutral and doped states of conjugated polymers also causes the concurrence of a number of important changes. The bipolaron states in a doped polymer are located in the gap between the conduction and valence bands (Figure 3). New electronic transitions are thus available to absorb photons often in the red to near infrared range. In the meantime, the band gap of the polymer is enlarged. As a result, poly(3,4-ethylenedioxythiophene) (PEDOT) appears blue whereas the doped PEDOT is sky blue or semi-transparent in thin films. The chromic change had been observed to study the electronic structure of conducting polymers. It is hard to determine who first suggested its use for electrochromic devices. As reviewed by Argun et al.,[18] conducting polymers are legitimate contenders for electrochromic display products thanks to the high contrast ratio, the availability of all colors due to the rich chemistry to fine tune the polymer's band gap, and the ease of processing polymer thin films.

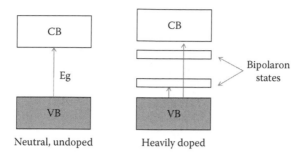

FIGURE 3   Band structure and electronic transitions in a conjugated polymer.

ELECTROMECHANICAL TRANSDUCTION

A volume change in a conjugated polymer during the insulator-conductor transition was not unexpected as the process involves ions in and out of the polymer. In 1989, Yoshino et al. reported large volume change in a poly(3-alkylthiophene) gel upon chemical doping with iodine or electrochemically.[19] In 1990, Baughman et al. first suggested the construction of electromechanical actuators based on the volume change.[20] Pei and Inganäs were able to achieve an "electrochemical artificial muscle" in 1991 with a polypyrrole/polyethylene bilayer beam that bent up to 90° upon electrochemical doping of polypyrrole.[21] In a series of publications, they showed that the volume change was determined by the nature of the ions moving in or out of polypyrrole, as illustrated in Figure 4. In PPy doped with perchlorate, perchlorate ion was ejected out of PPy when the polymer was dedoped, and the polymer's volume shrank. When the original dopant was bulky, like dodecylbenzenesulfonate that could not be ejected during dedoping, a cation from the solution inserted into the polymer to neutralize the idled DBS anion. With a modest-size dopant like tosylate, one may observe an initial volume expansion during the dedoping, followed by a large shrinkage because of salt draining and relaxation of the polymer matrix. The ionic transport and volume change could be affected by a number of other conditions such as swelling by the solvent and the porosity of the polymer. This simplified picture thus may alter.

The volume change in conducting polymers during electrochemical switching is generally small, less than 10%. Larger volume changes have been obtained in porous or gel-like polymers, although the cyclic stability can be low due to relaxation of the polymer matrix. Conducting polymer unimorph (bilayer beam) and bimorph actuators have been actively explored for several hydraulic and microelectromechanical systems.

$$1.\ PPy^+(ClO_4^-) + e^- \rightarrow PPy^0 + ClO_4^-$$

$$2.\ PPy^+(DBS^-) + e^- + Li^+ \rightarrow PPy^0(DBS^-\ Li^+)$$

$$2.\ PPy^+(TsO^-) + e^- + Li^+ \rightarrow PPy^0(TsO^-\ Li^+) \rightarrow PPy^0 + TsO^-\ Li^+$$

FIGURE 4   Illustration of the dedoping of polypyrrole (PPy) doped with various anions introduced during electroplolymerization.

LIGHT-EMITTING ELECTROCHEMICAL CELLS

The polymer light emitting diode reported by Burroughes et al.[22] in 1990 was arguably the most important development in the field of conjugated polymers since the discovery of conducting polyacetylene. The commercial potential of the device was quickly recognized and thus far, thousands of scientific papers have been published on polymer LEDs, bringing about new polymers and device structure for improved electroluminescent efficiency. It became clear since the early days that polymer LEDs required a high work function anode and a low work function cathode to minimize the barrier heights for electron and hole injections. The polymer must be used in the undoped state because the luminescence is readily quenched by doping. To overcome the charge injection barriers, Pei et al. proposed in 1994 the use of a blend of a conjugated polymer and a solid electrolyte instead of a pure conjugated polymer.[23] According to their idealized model, a bias voltage greater than the polymer's band gap causes the polymer to be symmetrically p-doped at the anode and n-doped at the cathode. The doped regions grow, and the growth frontiers will meet near the middle of the blend film. In the meantime, holes and electrons propagate and recombine, which counter the doping growth. The two processes arrive at equilibrium to form a dynamic p-i-n junction. In the i region, the polymer is undoped and fluorescent. Charge recombination in this region generates electroluminescence as in the LEDs. This device was conceived based on the electrochemical doping of conjugated polymers. It was thus termed a light-emitting electrochemical cell (LEC), despite the fact that the LEC is also a thin film solid-state device, like an LED. DeMello et al. argued that space charge accumulation at the electrode interfaces, not doping, was responsible for the formation of the Ohmic contacts.[24] This electrodynamic model could be valid in some material systems. However, the p-i-n junction in the LECs has been observed in a large number of polymer blends. The electronic and ionic mixed conductor system has been found to generate electroluminescence at high efficiency and low driving voltages. It was even shown that the p-i-n junction mechanism could turn a lousy polymer, with largely quenched fluorescence, to glow bright light: the polymer in the i-region would be dedoped and restore its pristine fluorescence. The mixed conductor system does pose additional challenges in materials preparation. The morphology or phase separation between the ionic regime and the electronic regime plays an important role in the electroluminescence. The ideal morphology for optimal electroluminescence has yet to be elucidated.

ELECTROCHEMICAL CIRCUITS

Conjugated polymers are also a strong contender for making electronic circuits. The motivations to use polymers are twofold: the possibility to shrink the circuit size to the molecular level and the printing of large-area circuits at extremely low-cost. Ionic carriers in the circuits add a new attribute to circuitry. In 1959, Lovrecek et al. reported an early example of an electrochemical rectifier based on polymers, polyvinyltolunesulfonic acid, and polyvinyltrimethylbenzylammonium hydroxide.[25] When the acid was biased positive and the hydroxide negative, the current carrying H^+ and OH^- ions were driven toward each other, met at the interface, and formed water. Therefore, current flow was easy. In the reverse bias, the mobile current carriers could not flow freely without electrolyzing water. Rectification ratios up to 85% were obtained.

Wrighton et al. described the first conducting polymer transistor in 1984.[26] The basic architecture of the transistor was a polypyrrole and three gold microelectrodes immersed in an electrolyte solution. The center gate electrode was applied a positive voltage to oxidize polypyrrole. As the polymer's impedance dropped, a current could flow between the left and right electrodes (source and drain). The switching time of such an electrochemical transistor was rather slow to drive a video-speed display. However, Berggren et al. recently showed that active matrix displays based on the electrochemical switching of PEDOT could be printed on paper.[27] All materials used were organic: the transistor channel utilized the insulator-conductor switching of PEDOT and the display element was based on the electrochromism of the same polymer. The printed displays were considered for fabricating electronic readers, like the Kindle® sold by Amazon.com, or electronic paper. These devices can tolerate a slow refreshable speed. This work manifested the versatile properties and low-cost fabrication of electronic devices containing ionic carriers.

FINAL REMARKS

The previously mentioned developments highlight the diversity and importance of iontronic organic materials and devices. The field is intertwined with so many other fields that a complete review of its history and all the important discoveries is almost impossible. I have provided merely my personal view on what occurred in the past that shaped the field as it is today. The authors who wrote the chapters of this book cover more history and exciting developments. The authors are active researchers who have shaped or are defining the field, and are making the strides to bring these materials into the marketplace. We hope this book can further the field, and provide a reference for and insights into iontronic organic electronic materials.

REFERENCES

1. D.M. Hercules, "Chemiluminescence resulting from electrochemically generated species," Science, 145, 808 (1964).
2. E.A. Chandross and F.I. Sonntag, "A novel chemiluminescent electron-transfer reaction," J. Am. Chem. Soc., 86, 3179 (1964).
3. S.S. Santhana and A.J. Bard, "Chemilunimescence of electrogenerated 9,10-Diphenylanthracene Anion radical," J. Am. Chem. Soc. 87, 139 (1965).
4. H.D. Abruna and Allen J. Bard, "Electrogenerated chemiluminescence. 40. A chemiluminescent polymer based on the tris(4-vinyl-4'-methyl-2,2'-bipyridyl)ruthenium(II) system," J. Am. Chem. Soc., 104, 2641 (1982).
5. H. Rudmann, S. Shimada, and M.F. Rubner, "Solid-state light-emitting devices based on the tris-chelated ruthenium(II) complex. 4. High-efficiency light-emitting devices based on derivatives of the tris(2,2'-bipyridyl) ruthenium(II) complex," J. Am. Chem. Soc., 124, 4918 (2002).
6. H. Kallmann and M. Pope, "Positive hole injection into organic crystals," J. Chem Phys. 32, 300 (1960).

7. M. Pope, H.P. Kallmann, and P. Magnante, "Electroluminescence in organic crystals," J. Chem. Phys. 38, 2042 (1963).
8. C. W. Tang and S. A. VanSlyke, "Organic electroluminescent diodes," Appl. Phys. Lett. 51, 913 (1987).
9. J. Ferraris, D. O. Cowan, V. Walatka, and J. H. Perlstein, "Electron transfer in a new highly conducting donor-acceptor complex," J. Am. Chem. Soc., 95, 948–949 (1973).
10. L. B. Coleman, M. J. Cohen, D. J. Sandman, F. G. Yamagishi, A. F. Garito, and A. J. Heeger, "Superconducting fluctuations and the Peierls instability in an organic solid," Solid State Commun. 12, 1125 (1973).
11. D. Jérôme, A. Mazaud, M. Ribault, and K. Bechgaard, "Superconductivity in a synthetic organic conductor $(TMTSF)_2PF_6$," Journal de Physique Lettres 41, L95 (1980).
12. V. V. Walatka, Jr., M. M. Labes, and J. H. Perlstein, "Polysulfur nitride—a one-dimensional chain with a metallic ground state," Phys. Rev. Lett. 31, 1139 (1973).
13. R. J. Nowak, H. B. Mark Jr. , A. G. MacDiarmid, and D. Weber, "Polymeric sulphur nitride: a new electrode material," J. Chem. Soc., Chem. Commun., 9 (1977).
14. W.A. Little, "Possibility of synthesizing an organic superconductor," Phys. Rev. 134, A1416 (1964).
15. H. Shirakawa, E. J. Louis, A. G. MacDiarmid, C. K. Chiang, and A. J. Heeger, "Synthesis of electrically conducting organic polymers: Halogen derivatives of polyacetylene, (CH) x," J. Chem. Soc., Chem. Commun., 578 (1977).
16. A. F. Diaz and J. I. Castillo, "Polymer electrode with variable conductivity: Polypyrrole," J. Chem. Soc., Chem. Commun., 397 (1980).
17. P. J. Nigrey, D. MacInnes, Jr., D. P. Nairns, A. G. MacDiarmid, and A. J. Heeger, "Lightweight rechargeable storage batteries using polyacetylene, (CH)x as the cathode-active material," J. Electrochem. Soc., 128, 1651 (1981).
18. A.A. Argun, P.H. Aubert, B.C. Thompson, I. Schwendeman, C.L. Gaupp, J. Hwang, N.J. Pinto, D.B. Tanner, A.G. MacDiarmid, and J.R. Reynolds, "Multicolored electrochromism in polymers: structures and devices," Chem. Mater. 16, 4401 (2004).
19. K. Yoshino, K. Nakao, M. Onoda, and R. Sugimoto, "Instability of conducting polymer, poly(3-alkylthiophene) gel with solvent, temperature, and doping and effect of alkyl chain length," Solid Slate Commun., 70, 609 (1989).
20. R.H. Baughman, L.W. Shacklette, R.L. Elsenbaumer, E.J. Plichta, and C. Becht, "Conducting polymer electromechanical actuators," in *Conjugated Polymeric Materials: Opportunities in Electronics, Optoelectronics, and Molecular Electronics*, NATO ASI Series E: Applied Sciences, 182:559 (Kluwer Academic Publishers, 1990).
21. Q. Pei and O. Inganas, "Conjugated polymers and the bending beam method: electric muscles and smart devices," Adv. Mater. 4, 277 (1992).
22. J. H. Burroughes, D. D. C. Bradley, A. R. Brown, R. N. Marks, K. Mackay, R. H. Friend, P. L. Burns, and A. B. Holmes, "Light-emitting diodes based on conjugated polymers," Nature 347, 539 (1990).
23. Q. Pei, G. Yu, C. Zhang, Y. Yang, and A.J. Heeger, "Polymer light-emitting electrochemical cells," Science, 269, 1086 (1995).
24. J.C. DeMello, N. Tessler, S.C. Graham, and R.H. Friend, "Ionic space-charge effects in polymer light-emitting diodes," Phys. Rev. B, 57, 12951 (1998).
25. B. Lovrecek, A. Despic, and J. O. M. Bockris, "Electrolytic junctions with rectifying properties," J. Phys. Chem., 63, 750 (1959).
26. H. S. White, G. P. Kittlesen, and M. S. Wrighton, "Chemical derivatization of an array of three gold microelectrodes with polypyrrole: Fabrication of a molecule-based transistor," J. Am. Chem. Soc., 106, 5375 (1984).
27. P. Andersson , D. Nilsson, P.-O. Svensson, M. Chen, A. Malmström, T. Remonen, T. Kugler, and M. Berggren, "Active matrix displays based on all-organic electrochemical smart pixels printed on paper," Adv Mater. 14, 1460 (2002).

The Editors

Magnus Berggren is the *Önnesjö* professor at Linköping University, Sweden, and guides the research activity of the organic electronics group. The 20 scientists of the organic electronics group aim at exploring electronic and optoelectronic functions of organic materials for paper electronics and bioelectronics applications. Dr. Berggren is the director of the Strategic Research Center for Organic Bioelectronics (OBOE) and is also the research manager at the Printed Electronics Arena (PEA). He collaborates closely with the Karolinska Institutet and several industries to explore novel organic electronics for drug delivery, tissue engineering, paper display, and sensor applications. In 1996, Berggren received his PhD in applied physics, from Linköping University, and started as a postdoctoral fellow at Bell Laboratories, Murray Hill, New Jersey. In 1997, he joined a team of investors and scientists to establish Thin Film Electronics AB (TFE), a company that develops printed organic memories. Berggren was the managing director for TFE until 1999. Then, he joined Acreo Institute to explore paper electronics and, in 2001, he became a professor of organic electronics at Linköping University.

Sue A. Carter received her BA in physics, chemistry, and mathematics from Kalamazoo College in Kalamazoo, Michigan, and her PhD in physical chemistry from the University of Chicago. She was a postdoctoral researcher at AT&T Bell Laboratories in Murray Hill, New Jersey, and a research fellow at IBM Almaden Research Center in San Jose, California. She is current a professor of physics at University of California, Santa Cruz. Over the last 15 years, her research has focused on the electronic, magnetic, thermal, and optical properties of inorganic, organic, and biological materials. Currently, she is studying the application of nanostructured materials to next generation energy technologies, including solid-state lighting, solar cells, and electrochemical cells. She has published over 50 articles in these areas, has organized several conferences for the APS, MRS, and SPIE, and has given dozens of invited talks in the areas of organic electronics, metal oxides, and renewable energy technologies. She served as the chief technical advisor for Add-vision, charting their technology pathway for developing fully printed OLED lamps. She is currently on the scientific advisory board for Solexant. She is a recipient of a 1996 Packard Fellowship.

Janelle Leger is an assistant professor at Western Washington University in the Department of Physics and Astronomy, the Department of Chemistry, and the Advanced Materials Science and Engineering Center. Her research group explores organic and hybrid electronic and optoelectronic devices, as well as structures for subwavelength optics. Leger was an NSF Discovery Corps Postdoctoral scholar at the University of Washington Department of Chemistry from 2005 to 2008, after receiving her PhD from the University of California at Santa Cruz Department of Physics. She also spent one year working as a research scientist at Add-Vision, Inc.

List of Contributors

Amanda Norell Bader
Western Washington University
Bellingham, WA

Magnus Berggren
Linköping University
Norrköping, Sweden

Sue Carter
University of California–Santa Cruz
Santa Cruz, CA

Fabio Cicoira
Cornell University
Ithaca, NY

Xavier Crispin
Linköping University
Norrköping, Sweden

Ludvig Edman
Umeå University
Umeå, Sweden

Peter Anderson Ersman
Acreo AB
Norrköping, Sweden

Lars-Olov Hennerdal
Acreo AB
Norrköping, Sweden

Lars Herlogsson
Linköping University
Norrköping, Sweden

Olle Inganäs
Linköping University
Norrköping, Sweden

Edwin Jager
Linköping University
Norrköping, Sweden

Oscar Larsson
Linköping University
Norrköping, Sweden

Janelle Leger
Western Washington University
Bellingham, WA

Yongfang Li
Institute of Chemistry
Chinese Academy of Sciences
Beijing, China

Mark Lonergan
University of Oregon
Eugene, OR

George Malliaras
Centre Microélectronique de Provence
Ecole Nationale Supérieure des Mines
   de Saint Etienne
Gardanne, France

David Nilsson
Acreo AB
Norrköping, Sweden

Qibing Pei
University of California, Los Angeles
Los Angeles, CA

Minh Chau Pham
University of Paris
Paris, France

Benoît Piro
University of Paris
Paris, France

Stephen Robinson
University of Oregon
Eugene, OR

Elias Said
Linköping University
Norrköping, Sweden

Nayoung Shim
Cornell University
Ithaca, NY

David Stay
University of Oregon
Eugene, OR

Payman Tehrani
Linköping University
Norrköping, Sweden

Sang Yang
Cornell University
Ithaca, NY

1 Electrochemistry of Conjugated Polymers

Yongfang Li and Qibing Pei

CONTENTS

1.1   ABSTRACT

Conjugated polymers, such as polyacetylene and polypyrrole, have attracted great attention since the discovery in 1977 of high electrical conductivity in doped polyacetylene. These polymers also exhibit important semiconductor properties at the intrinsic, undoped state. Electrochemistry has played a key role in the studies of conjugated polymers. In this chapter, we focus our attention on the electrochemistry of conjugated polymers, including electropolymerization to prepare the polymers directly from their monomers, the electrochemical properties of conducting polymers, and the electrochemical measurements of the HOMO and LUMO energy levels of conjugated polymers.

Conjugated polymers, both as semiconductors at the intrinsic state and as metallic conductors when heavily doped, have attracted broad interests since the discovery of conducting polyacetylene by Shirakawa, MacDiarmid, Heeger, et al. in 1977.[1] A variety of potential applications have been studied based on the novel physicochemical properties of these polymers. The possible application fields of conducting polymers and semiconducting polymers include electrode materials of batteries, electrochromics, modified electrodes, enzyme electrodes, solid capacitors, polymer light emitting diodes, and polymer solar cells. Most of the applications are related to the electrochemical properties of the conjugated polymers. Meanwhile, electrochemical polymerization is the main preparation method for a number of widely used conducting polymer films such as polypyrrole and polythiophene. Therefore, electrochemical polymerization and electrochemical properties of conducting polymers have been among the main research subjects of conducting polymers since the early 1980s. This chapter reviews the scientific understanding and important scientific discoveries made on the electrochemistry of conjugated polymers.

1.2   ELECTROCHEMICAL PREPARATION OF CONDUCTING POLYMERS

Among various synthetic methods of conducting polymers, the electrochemical polymerization played an important role in the preparation of conducting polypyrrole, polyaniline, and polythiophene because the electrochemical approach has the advantage of one-step deposition of conducting polymer films onto a metal electrode surface. The thickness of the conducting polymer films can be controlled precisely by controlling the amount of charges consumed in the electropolymerization. Moreover, the properties of the conducting polymer films produced electrochemically can be regulated easily by changing the counter-anions in the electrolyte solutions. In addition to the preparation of self-standing conducting polymer films, the electropolymerization is also a preferable method in the preparation of polymer modified electrode, enzyme electrode (containing immobilized enzyme in the conducting polymers), and polymer electrode for electrochromic displays.

The electrochemical preparation of conducting polymers is usually performed by the oxidative polymerization of their corresponding monomers. Three methods, constant current, constant potential, and cyclic potential scanning, are commonly utilized. Two electrodes (an anode and a cathode, in the constant current method) or three electrodes (an anode, a cathode, and a reference electrode, in the potential-controlled methods) in an electrolyte solution of the monomers are needed for the electropolymerization. The anode, on which the conducting polymer is deposited, could be Pt, Au, carbon, stainless steel, or indium-tin oxide (ITO) conducting glass. There are many factors influencing the electropolymerization processes, such as solvents, supporting electrolyte salts, concentration of the monomers, pH value of the electrolyte solutions, polymerization potential, current, temperature, etc. Polymerization potential of the monomers is the key factor to affect the electropolymerization. Actually, most of the conducting polymers can be prepared by the electropolymerization of its corresponding monomers, but the oxidation polymerization potential is different for different monomers. Table 1.1 shows the polymerization potentials of several important monomers for the synthesis of conducting polymers. The lower the oxidation polymerization potential is, the easier the electropolymerization of the monomers will be. Obviously, the electropolymerization of pyrrole is the easiest among the monomers. The electrochemical oxidation potentials of pyrrole and aniline are lower than the oxidation decomposition potential of acidic water; therefore, the electropolymerization of pyrrole and aniline can be performed in acidic aqueous solutions. However, the electropolymerization potential of thiophene is higher than 1.5 V vs. SCE, so the electropolymerization of thiophene has to be carried out in a nonaqueous solution.

The structure and conductivity of the as-prepared conducting polymer films are strongly dependent on the electropolymerization conditions, including electrolyte composition, polymerization potential, current, and temperature. The correlation is determined by the electropolymerization mechanism. The studies on the electropolymerization processes have mainly focused on the elucidation of the electropolymerization mechanism and on the optimization of the polymerization conditions.

The most important conducting polymers synthesized by electrochemical method are polypyrrole, polythiophene, and polyaniline. These three polymers are air-stable, highly conductive, and promising for applications in many aspects. The electrochemical preparations of these conducting polymers are described in the following.

TABLE 1.1

Oxidative Polymerization Potential of Monomers and the Conductivity of the As-Prepared Conducting Polymers[2]

Monomer	Polymerization Potential (V vs. SCE)	Conductivity of the As-Prepared Polymers (S/cm)
Pyrrole	0.7	30–100
Aniline	0.8	1–20
Indole	0.9	0.005–0.01
Azulene	0.9	0.01–0.1
Thiophene	1.6	10–100

High conductivity and strong mechanical strength are key properties to optimize in the preparation of conducting polymer films. Much of the following discussions on polymerization conditions are focused on how to improve these properties and the corresponding structures.

1.2.1 ELECTROCHEMICAL PREPARATION OF POLYPYRROLE

Polypyrrole (PPy) is one of the most stable and environmentally friendly conducting polymers. It is the first conducting polymer produced by electropolymerization. As early as 1968, Dall'Olio et al. prepared conducting PPy film with conductivity of 8 S·cm^{-1} by electrochemical polymerization,[3] but this work did not draw attention at that time. Ten years later, Diaz et al. synthesized conducting PPy films with conductivity reaching approximately 100 S/cm by electrochemical oxidation polymerization of PPy in acetonitrile electrolyte solution.[4] Diaz's work attracted immediate attention among the broad search for new conducting polymers after the discovery of doped polyacetylene.[1] The electropolymerization of PPy can be done both in organic and in aqueous solutions thanks to its low oxidation potential.

1.2.1.1 Effect of Electrolyte Solution on the Electropolymerization

Proper control of the electropolymerization conditions is crucial in the preparation of high quality PPy films. The electropolymerization is generally performed in an electrolyte solution containing solvent, electrolyte salt, and the monomer. In the following, the effects of the electrolyte anions, solvent, concentration of the electrolyte salt, and pH value of an aqueous solution on the polymerization reaction are discussed, respectively.

Table 1.2 shows the effect of anions of the supporting electrolyte in organic solutions.[5] The conductivity (σ) of the as-prepared PPy films strongly depends on the nature of the anions. It varies from the order of 10^{-1} S/cm to the order of 10^2 S/cm for different counter anions. Qian and Qiu studied the electropolymerization of PPy in aqueous solutions, and found similar strong dependence of conductivity of PPy films on the solution anions, as shown in Table 1.3.[6] In the aqueous solutions, flexible PPy films with σ higher than 100 S/cm are produced with the surfactant anions such as tosylate, benzene sulphonate, etc. Warren and Anderson correlated the anion effect with their basicity and found that the stronger the conjugated acids of the anions, the higher the conductivity of the PPy films prepared from the anion-containing solution.[7] In general, larger sized and surfactant anions of the strong acids are favorable for the preparation of high quality PPy films.

The concentration of the electrolyte anions also plays an important role.[8–12] Too low concentration of the anions will lead to poor PPy films. The concentration of the anions should be at least 0.1 mol/L, and the commonly used concentration is in the range of 0.1 ~ 1.0 mol/L. The concentration of PPy monomer is usually 0.1 mol/L.

Different electrolyte anions also lead to various surface morphologies of the as-prepared PPy films. For instance, the surface of the PPy film prepared from PF_6^-

TABLE 1.2

Effect of Electrolyte Anions on the Properties of As-Prepared PPy Films

	Properties of the As-Prepared PPy Films		
Supporting Electrolyte	Doping Degree of the Counter Anions	Density/ $g \times cm^{-3}$	Conductivity/ $S \times cm^{-1}$
BF_4^-, PF_6^-, AsF_6^-	0.25 ~ 0.32	1.48	30 ~ 100
ClO_4^-	0.30	1.51	60 ~ 200
HSO_4^-	0.30	1.58	0.3
$CF_3SO_3^-$	0.31	1.48	0.3 ~ 1
TsO^-	0.32	1.37	20 ~ 100
CF_3COO^-	0.25	1.45	12

Note: The electropolymerization was performed in 0.06 M PPy, 0.1 M supporting electrolyte (tetraethylammonium or tetrabutylammonium salts) acetonitrile solution with 1% water.

TABLE 1.3

Effect of the Electrolyte Anions in Aqueous Solutions on the Conductivity and Density of the As-Prepared Conducting PPy Films [6]

	Properties of the As-Prepared PPy Films		
Supporting Electrolytes	Conductivity/ $S \times cm^{-1}$	True Density/ $g \times cm^{-3}$	Apparent Density/ $g \times cm^{-3}$
$HClO_4$	20 ~ 30	1.575	0.52
$NaClO_4$	—	1.558	0.36
$NaBF_6$	12	1.540	0.38
KPF_6	12	1.549	0.46
$TsOH$	60 ~ 200	1.368	1.24
$TsONa$	60 ~ 200	1.364	1.25
$NaNO_3$	4 ~ 30	1.516	1.25
H_2SO_4	10	1.532	1.26

containing aqueous solution shows bowl-shaped microstructures, while that obtained from the ClO_4^- aqueous solution displays cauliflower morphology.[6]

The effect of solvent on the electropolymerization was found to be related to the donor number (DN) of the solvent.[13] The lower the DN value of the solvent, the higher the conductivity of the as-prepared PPy films, as shown in Table 1.4. The solvent effect can be explained from the cation radical coupling mechanism of the electropolymerization. High DN value of the solvent means high nucleophilicity or strong basicity. The solvent molecules with high DN value attack the cation radical formed by oxidation of PPy, which hampers the electropolymerization.

TABLE 1.4

Effect of Solvent on the Conductivity of As-Prepared PPy Films

Solvents	DN	Conductivity of the As-Prepared PPy films (S/cm)		
		$PPy(BF_4^-)$	$PPy(ClO_4^-)$	$PPy(TsO^-)$
DMSO	29.8	7×10^{-6}		
DMF	26.6	1×10^{-4}	5×10^{-4}	0.008
TBP	23.7		1	
TMP	23.0	1.0	20	0.09
THF	20.0		31	
H_2O	18.0	8.4	34	79
PC	15.1	67	55	90
CH_3NO_2	2.7	69	56	

Water is a special solvent in comparison with organic solvents; its acidity can be regulated by changing pH values. The optimal pH value of the aqueous solutions for pyrrole electropolymerization is between pH 2 and pH 5.5.[14] In a basic aqueous solution, conducting PPy cannot be produced.[15]

In addition to the solvent and supporting electrolyte, small amounts of additives in the solutions can sometimes affect the electropolymerization. Surfactants have been proven to be effective in improving the quality of the PPy films.[16–19] With nonionic surfactant nonylphenol polyethyleneoxy ether (10) as an additive in the TsONa aqueous solution, the tensile strength of the PPy film produced from the solution reached 127 MPa, which is five times higher than that of the PPy film produced from the solution without the surfactant additive.[18] The use of additives can also have a significant effect on the conductivity and surface smoothness of the as-prepared PPy films, as shown in Table 1.5.[19] These effects may be due to the absorption of the surfactant molecules on the surface of the working electrode. The modification of the interfacial structure between the electrode and the solution would affect the deposition process of PPy and the quality of the resulting PPy films.

1.2.1.2 Effects of Potential, Current, and Temperature on the Electropolymerization

As mentioned previously, the electropolymerization can be performed with a potential-controlled method (constant potential or potential scanning) or current-controlled method (constant current). In the potential-controlled method, the potential should be controlled no higher than 0.75 V vs. SCE. The electropolymerization is usually carried out at 0.65 ~ 0.70 V vs. SCE. The potential over 0.8 V vs. SCE will result in overoxidation and degradation of the deposited PPy films.[20]

Constant current polymerization is widely used in the preparation of PPy films because of its convenience in operation and in controlling the total amount of polymerization charges from which the thickness of the PPy films can be controlled.

TABLE 1.5

Effect of Nonionic Surfactant Additives on the Conductivity and Mechanical Properties of the Electropolymerized PPy Films[19]

Additives	Surface Morphology	Conductivity (S/cm)	Tensile Strength (M Pa)	Elongation at Break (%)
Without additive	Rough	94.0	24.0	5.0
OP$_4$	Smooth	48.3	25.4	6.3
OP$_7$	Very smooth	84..1	60.8	16
OP$_{10}$	Very smooth	92.2	68.4	18
OP$_{15}$	Very smooth	89.8	64.3	16
OP$_{21}$	Very smooth	113.2	67.0	20
A-20	Smooth	126.6	66.3	17
SE-10	Smooth	76.7	58.7	8.0

Note: The PPy films were prepared at 1 mA/cm² at 15°C from 0.1 mol/L pyrrole, 1 mol/L TsONa neutral aqueous solutions adding 0.01 mol/L various additives.

Maddison et al.[21] studied the effect of current density and found that the PPy film with highest conductivity was obtained at 2.8 mA/cm². The appropriate current density is dependent on the other conditions of the polymerization solutions. Good PPy films are usually produced at a current density of 1 ~ 2 mA/cm².

Temperature also plays a role in electropolymerization.[14] In general, high quality PPy films with longer conjugation chains, less defects, and higher conductivity could be obtained at lower temperatures (lower than 20°C).[8,22–24] At higher temperatures, defect structures of PPy are easily formed and result in decrease of its conductivity. For the polymerization in a TsO⁻ aqueous solution, the proper temperature is in the range of 1 ~ 20°C.[21]

1.2.1.3 Mechanism of the Electropolymerization of Pyrrole

The electrochemical preparation of PPy is carried out by the oxidation of pyrrole in an electrolyte solution. Naturally, one can propose a cation-radical mechanism.[25] According to this mechanism, at first, pyrrole monomers are oxidized into cation radicals on the anode. Two cation radicals couple together to form a dimer, with concurrent expelling of two protons. The dimer is then oxidized into its cation radical, more easily than the monomer because of its lower oxidation potential. The cation radical of the dimer will couple with a cation radical of the monomer to form a trimer. The trimer is readily oxidized and coupled to a radical cation to form a tetramer. The polymer chain grows in length rapidly. The PPy deposition therefore follows a nucleation and growth mechanism. The PPy films produced by electropolymerization are oxidized into their *p*-doped (conductive) state because the *p*-doping potential of the polymer is much lower than the oxidation potential of the monomer.

The cation-radical mechanism can explain successfully the effect of solvent and solution anions on the electropolymerization. However, it fails to account for the effect of pH values of the electrolyte solutions. Qian et al.[15] modified the cation-radical mechanism with a pre-protonation assumption. In this mechanism, pyrrole monomer is protonated on its β-carbon, resulting in a change of its electronic structure, which benefits the formation of a pyrrole cation radical. The polymerization still follows the cation-radical mechanism. In essence, the protons behave like a catalyst in the polymerization reaction. This mechanism explains why a low pH value is favorable for the electropolymerization of pyrrole.

However, there is a serious problem with the cation-radical polymerization mechanism, which is the Columbic repulsion between two cation radicals when they come in close contact with each other to couple into a longer chain. In addition, the mechanism did not consider the effect of solution anions on the electropolymerization. Actually, the concentration of the solution anions significantly affects the oxidative polymerization rate.[26] Morphology and chain structure of PPy films are also affected by the solution anions. It was also found that the conductivity difference of the PPy films with different counter-anions is not from the interaction between the PPy chain and the counter-anions, but from the different chain structures formed during its polymerization.[27] Therefore, the solution anions should take part in the electropolymerization processes. The experimental results of competitive doping of two different anions into PPy during pyrrole polymerization support the anion-participated mechanism.[28,29] Based on Zotti's ion-pairing mechanism,[11] the following anion-participated mechanism[26] was proposed for the electropolymerization of pyrrole:

$$(H-N-H) \underset{k_{-1}}{\overset{k_1}{\rightleftarrows}} (H-N-H)_{ad} \tag{1.1}$$

$$(H-N-H)_{ad} - e + A^- \xrightarrow{k_2} A^-(H-N-H)^{+\bullet}_{ad} \tag{1.2}$$

$$2[A^-(H-N-H)^{+\bullet}_{ad}] \xrightarrow{k_3} (H-N-N-H) + 2A^- + 2H^+ \tag{1.3}$$

$$(H-N-N-H) - e + A^- \xrightarrow{k_4} A^-(H-N-N-H)^{+\bullet}_{ad} \tag{1.4}$$

$$A^-(H-N-N-H)^{+\bullet}_{ad} + A^-(H-N-H)^{+\bullet}_{ad} \xrightarrow{k_5} (H-N-N-N-H) + 2A^- + 2H^+ \tag{1.5}$$

$$(H-N-N-N-H) - e + A^- \xrightarrow{k_6} (H-N-N-N-H)^+ A^- \tag{1.6}$$

In the above reaction equations, H-N-H represents a pyrrole molecule and the two H's represent the two α-H of pyrrole.[30] The mechanism in Reactions (1.2) and (1.4) is similar to the ion-pairing mechanism proposed by Zotti et al.[11] The complex of the ion-pairing reduced the electrostatic repulsion between two cation radicals, which benefits the coupling of the two cation radicals for the chain growth.

The anions-participated electropolymerization mechanism can explain all the phenomena related to the effects of polymerization conditions on the electropolymerization processes observed in the experiments. In addition, the Columbic repulsion between two cation radicals is overcome through ion pairing of the cations with anions. The kinetic equation deduced from the mechanism also agrees with the experimental results, which will be discussed in the following section.

1.2.1.4 Kinetic Equation of the Electropolymerization of Pyrrole

Based on the anions-participated electropolymerization mechanism, the kinetic equation of the pyrrole electropolymerization has been derived.[26]

The following assumptions are made:

1. The adsorption of pyrrole molecules on the anode is in equilibrium during the electrode reaction.

$$Py ==== (Py)_{ad} \tag{1.7}$$

2. The solution anions adsorb on the anode competitively with the pyrrole molecules.
3. The rate-limiting step of the polymerization is the oxidation reaction of pyrrole molecules into the complex of pyrrole cation radical and the solution anion:

$$(Py)_{ad} - e^- + A^- \xrightarrow{k_2} (Py)_{ad}^{+\cdot} A^- \tag{1.8}$$

Assumption 3 is reasonable because the oxidation potential of the pyrrole molecules is the highest in comparison with that of its dimer, oligomer, or polymer. The polymerization of pyrrole is performed at a positive potential, so the competitive adsorption of the solution anions (Assumption 2) should occur on the electrode with positive potential.

Based on Assumptions 1 and 2, the concentration of the adsorbed pyrrole on the electrode can be expressed as Equation (1.9) according to the Langmuir adsorption equation:

$$x = \frac{K_p[Py]}{1 + K_p[Py] + K_a[A^-]} \tag{1.9}$$

In addition, according to Assumption 3,

$$I \propto r_2 = k_2 x[A^-] \tag{1.10}$$

Then:

$$I = K'k_2 \frac{K_p[Py][A^-]}{1 + K_p[Py] + K_a[A^-]} \tag{1.11}$$

or:

$$I = \frac{K[Py][A^-]}{1 + K_p[Py] + K_a[A^-]} \tag{1.12}$$

where $K = K'k_2K_p$

Two extreme cases can be deduced from Equation (1.12):

1. For low concentration solutions of the anions—When the concentration of the anions is very low, so that $K_a[A^-] \ll 1$, the $K_a[A^-]$ item in the denominator of Equation (1.12) is negligible. The equation becomes

$$I \approx \frac{K[Py][A^-]}{1 + K_p[Py]} \tag{1.13}$$

Therefore, the current or reaction rate will be proportional to the concentration of the solution anions.

2. For high concentration solutions of the anions—When the concentration of the anions is very high and the concentration of pyrrole molecules is low, so that $K_a[A^-] \gg 1 + K_p[Py]$, the first and second items in the denominator of Equation (1.12) is negligible. The equation becomes

$$I \approx \frac{K[Py][A^-]}{K_a[A^-]} = \frac{K[Py]}{K_a} \tag{1.14}$$

Therefore, the current or reaction rate will be independent of the concentration of the solution anions.

The kinetic equations (1.12), (1.13), and (1.14) agree with the experimental results,[26] which further supports the anions-participated mechanism of the pyrrole electropolymerization.

1.2.2 ELECTROCHEMICAL PREPARATION OF POLYTHIOPHENE

The electrochemical preparation process of conducting polythiophene (PTh) is carried out by the oxidation polymerization of thiophene, similar to that of PPy. However, the polymerization potential (approximately 1.6 V vs. SCE) of thiophene monomers is much higher than that (approximately 0.7 V vs. SCE) of pyrrole, as shown in Table 1.1. The higher oxidative polymerization potential of thiophene requires that the polymerization has to be performed in nonaqueous solution because water molecules will be electrolyzed before the oxidative polymerization of thiophene in an aqueous solution. Inert electrode and electrolyte should be used to avoid the oxidation reaction of them at the high polymerization potential of thiophene. In addition, the overoxidation degradation of the as-prepared PTh films should also be avoided during the electrochemical polymerization. Obviously, decreasing the polymerization potential is beneficial to the preparation of high-quality PTh films. Therefore, many efforts have been devoted to decreasing the polymerization potential. Diaz et al.[31] and Garnier et al.[32] studied the effect of substituents of thiophene on the polymerization potential. A part of their results is listed in Table 1.6. It can be seen that bithiophene and thiophene derivatives with electron-donating group have lower polymerization potentials than that of thiophene. The electropolymerization of thiophene oligomers and thiophene derivatives have attracted considerable attention in the electrochemical preparation of polythiophene.

1.2.2.1 Electropolymerization of Thiophene Monomers

In 1982, Tourillon and Garnier[2] prepared PTh electrochemically in a solution of CH_3CN containing 0.1 M $(Bu)_4NClO_4$, 0.01 M thiophene, and approximately 0.01 M water. The solution was pretreated by bubbling with argon for 15 min. Electropolymerization at 1.6 V vs. SCE on a Pt anode produced a PTh film with the conductivity of 10 to 100 S/cm. Structural analysis indicates that the polymer chains

TABLE 1.6
The Oxidative Polymerization Potential of Thiophene and Its Derivatives[32]

Monomers	Oxidative Polymerization Potential (V vs. SCE)
Thiophene (T)	1.65
2,2'-bithiophene (2,2'-bT)	1.20
3-methyl thiophene (3-MeT)	1.35
3-bromothiophene (3-BrT)	1.85
3,4-dibromo thiophene (3,4-BrT)	2.00
3,4-dimethyl thiophene (3,4-MeT)	1.25
3,4-methyl ethyl thiophene (3,4-MeEtT)	1.26
3,4-diethyl thiophene (3,4-EtT)	1.23
3-thiomethyl thiophene (3-CH₃ST)	1.30

are connected by an α–α connection. Waltman et al[31] found that thiophene derivatives substituted at the α-position could not be electropolymerized, which supports the α–α linkage of PTh chains.

Tourillon and Garnier[2] also found that the pretreatment of the polymerization solution by Ar bubbling is very important. Without the Ar pretreatment to remove oxygen from the electrolyte solution, the conductivity of the resulting PTh films was only approximately 0.1 S/cm, 2 to 3 orders of magnitude lower than that of the PTh films prepared with the Ar pretreatment. Because the polymerization potential of thiophene is quite high, the PTh films deposited on the anode are overoxidized in the presence of oxygen and water during the electropolymerization. The existence of C=O in the IR spectra of the PTh film as-prepared without the Ar pretreatment confirmed the occurrence of overoxidation.[2]

The effects of various other polymerization conditions on the electropolymerization of thiophene were also studied. Tanaka et al.[33] studied the effect of anode metal materials on electropolymerization and found that PTh films can be deposited on Pt, Au, Cr, Ni, ITO, etc. but there is no PTh formed when Cu, Ag, Pb, or Zn were used as the anode. This is because Cu, Ag, Pb, and Zn are electrolyzed and dissolved at the polymerization potential of thiophene. The dielectric constant of the solvent in the polymerization solutions also affects the electropolymerization processes. Table 1.7 shows the results from several important organic solvents. Obviously, the electropolymerization can only be carried out in solvents with higher dielectric constants.

The concentration of the thiophene monomer is also important for polymerization.[34] The onset potential of thiophene electropolymerization decreases with the increase of thiophene concentration. Usually, the concentration of the thiophene monomer should be no lower than 0.1 mol/L. Too low thiophene concentration will result in overoxidation of the deposited PTh film.[35]

Anions with oxidation potential lower than 1.7 V vs. SCE cannot be used in the electrolyte solutions because the oxidation of the anions will alter the oxidative

TABLE 1.7

The Effect of Solvent on the Electropolymerization of Thiophene in a 0.1 M Bu_4NClO_4 Solution

Solvents	Dielectric Constant	Lowest Thiophene Concentration Required for the Electropolymerization (mol/L)
Acetonitrile	37.50	0.1
Nitrobenzene	35.70	0.1
Benzonitrile	25.20	0.4
Chlorobenzene	5.71	No polymerization
Anisole	4.33	No polymerization

TABLE 1.8

Mechanical Properties of As-Prepared and Neutral Polythiophene Films[a]

Samples	Polymerization Current Density (mA/cm²)	Modulus (GPa)	Strength (MPa)
As-polymerized films	0.7	2.6	74
	3.0	2.1	60
	5.0	1.8	35
Neutral films	0.7	3.3	82
	3.0	2.6	67
	3.3	2.0	30

[a] The electropolymerization was performed in 0.3 M thiophene, 0.03 M Bu4NClO4 in dehydrated nitrobenzene at 5°C.

---

polymerization of thiophene.[36] Therefore, salts of I⁻, Br⁻, and Cl⁻ are not suitable as the supporting electrolyte.

Ito et al.[37] studied the effect of polymerization current on the mechanical properties of PTh films. They found that the strength of the PTh films increased with the decrease of the current (see Table 1.8). In addition, the surface of the PTh films is smoother from lower current polymerization.

Similar to the electropolymerization of PPy, lower temperature favors the formation of higher quality PTh films with higher conductivity and mechanical flexibility.[38–40] For the PTh films produced at high temperatures (such as 40°C), the effective π-conjugation length in the PTh chains is rather short. The polymer contains cross-linked short chain fragments of PTh.[38] Osawa et al.[40] studied the effects of temperature and magnetic field on the electropolymerization and found that under a magnetic field PTh chains align along the electrode surface if the polymerization is carried out at low temperatures (5°C). There is no molecular alignment if the polymerization temperature is higher than 5°C and the current is larger than 1 mA/cm². Solvent and polymerization current also affect the molecular orientation. The direction of the magnetic field is also very important for the molecular orientation; the magnetic field parallel to the electrode surface helps the chain orientation, while that perpendicular to the electrode surface breaks the alignment.[40]

Shi et al. found that the polymerization potential of thiophene can be decreased by approximately 0.6 V by using the electrolyte solution of boron trifluoride/ethylene ether (BFEE),[41] which is very important for the preparation of high-quality conducting PTh films. Li et al. further studied the effect of the mixed solvent of BFEE and acetonitrile, and found that the PTh film prepared from the mixed solution of 90% BFEE and 10% acetonitrile possessed the highest conductivity.[42]

The electropolymerization of thiophene follows a mechanism of cation radical–cation radical coupling,[34,43] nucleation, and three-dimensional growth,[44] which is very similar to that of the pyrrole electropolymerization. Solution anions should also participate in the electropolymerization process of thiophene like that of pyrrole.[26]

1.2.2.2   Electropolymerization of Thiophene Oligomers

As mentioned previously, the polymerization potential of bithiophene is 1.20 vs. SCE, decreased by 0.45 V in comparison with that of thiophene. The lower polymerization potential makes the bithiophene easy to be electropolymerized. Therefore, many researchers prepare PTh by the electropolymerization of bithiophene.[45–48] The polymerization potential of terthiophene decreases further.[49–51] Poly(terthiophene) was found to be partially crystalline (d = 4.7, 4.0, 3.3 Å).[50] However, the molecular weight and conductivity of the PTh films prepared from bithiophene and terthiohene are lower than those of the PTh films obtained from thiophene.

Hu et al.[45] found that by using $HClO_4$ as the electrolyte in a mixture of acetonitrile and water, the electropolymerization potential of bithiophene can be decreased further. In the 0.02 mol/L bithiophene, 1.0 mol/L $HClO_4$ mixed water and acetonitrile solution, the polymerization potential of bithiophene depends on the ratio of water to acetonitrile, being 0.62, 0.75, and 0.80 V vs. Ag/AgCl at the water:acetonitrile volume ratios of 2:1, 1:1, and 1:4, respectively.[45] Protons in the solution may play a catalytic role in the solution system. In a surfactant aqueous solution, PTh film can be deposited on an iron electrode by the electropolymerization of bithiophene.[45] The surfactant in the aqueous solution is helpful to increase the solubility of bithiophene and to decrease the polymerization potential.

Zotti et al.[52] studied the kinetics and mechanism of the electropolymerization of thiophene oligomers. The results show that the reaction rate of the polymerization depends on the length of monomer. The kinetics is second order. The coupling reaction has high activation enthalpy and negative activation entropy. The coupling rate decreases as the oligomer length is increased.[52] The smaller molecular weight of the as-prepared PTh from thiophene oligomers may relate to the slow polymerization rate.

1.2.2.3   Electropolymerization of Thiophene Derivatives

Soluble polythiophene derivatives have attracted considerable interest in recent years because of their promising applications in solution-processable optoelectronic devices. The PTh derivative can be prepared by the electropolymerization of its corresponding thiophene derivative monomers. Moreover, the polymerization potentials of thiophene derivatives with alkyl or alkyloxy substituent groups, which are the most interesting thiophene derivatives, decreased in comparison with that of thiophene (see Table 1.6), and the conductivity of the PTh derivatives is higher than that of unsubstituted PTh. 3-Methoxythiophene can be electropolymerized in an aqueous solution, thanks to its lower polymerization potential with the substitution of strong electron-donating methoxy group.[53] Surfactant was used in the aqueous solution to increase the solubility of thiophene derivatives.[53]

Higgins et al.[54] performed the electropolymerization of benzo[c]thiophene in an acetonitrile solution and found that its polymerization potential is as low as 0.7 ~ 0.8 V vs. SCE. The deposited polymers show interesting electrochemical properties.[53] The electropolymerization of 3-(1-naphthylthiophene) was also carried out in 0.1 mol/L Bu_4NBF_4/acetonitrile solution,[55] its polymerization potential is 1.1 ~ 1.2 V vs. Ag/Ag+, lower than that of thiophene but higher than the potential of phenyl substituted thiophene (0.991 V). Lower polymerization potential of naphthylthiophene should be expected because the electron donating ability of naphthyl is stronger than that of phenyl. The higher potential results from sterical hindrance of the bulky naphthyl group. The alkylester-substituent on thiophene also decreased the polymerization potential (1.55 V vs. SCE[56]) of thiophene.

Electron-withdrawing substituents, such as Br, CN-, etc., on the thiophene ring shift the polymerization potential to higher potential, as shown in Table 1.7. But by using some special solution system, such as BFEE, poly(3-brominethiophene) can be prepared by the electropolymerization of 3-brominethiophene at 1.61 V vs. SCE.[57] When BFEE is mixed with 30% trifluoroacetic acid (TFA), the polymerization potential of chloride substituted thiophene decreased further in comparison with that in pure BFEE solution (from 1.54 V vs. SCE in BFEE to 1.16 V vs. SCE in BFEE with 30% TFA).[58]

Poly(3,4-ethylenedioxythiophene) (PEDOT) has drawn much attention recently because of its applications as a modifying layer on an ITO electrode for polymer LED and polymer solar cell devices, transparent electrode materials, antistatic painting materials, and in solid capacitors. PEDOT can also be prepared by the electropolymerization of its monomer EDOT. The polymerization potential of EDOT is 1.49 V vs. SCE, which is lower than that of thiophene. By using a dimer of EDOT as the monomer, the potential can be decreased further to 0.84 V vs. SCE.[59]

1.2.3 ELECTROCHEMICAL PREPARATION OF POLYANILINE

The potential of oxidative polymerization of aniline is approximately 0.8 V vs. SCE, and an acidic medium is crucial for the electropolymerization. In addition, the electrochemically deposited polyaniline (PANi) layer on the electrode cannot be peeled off as flexible films like PPy and PTh films prepared by electropolymerization. Hence, the electrochemical preparation of PANi is mainly for deposition of a PANi modified electrode for basic electrochemical studies and for applications in electrocatalysis, sensors, and electrode materials in lithium batteries.

1.2.3.1 Electropolymerization in Acidic Aqueous Solutions

The electropolymerization of aniline is usually performed in an acidic aqueous solution. Commonly used solutions contain 0.1 mol/L aniline and 1 mol/L H_2SO_4, HCl, $HClO_4$, or HBF_4 in water.[60–69]

The conductivity of the PANi prepared in the acidic aqueous solutions is usually on the order of 10^{-1} to 10^0 S/cm. The morphologies of the as-prepared PANi depend on the electropolymerization conditions. With a constant current method (5 mA/cm^2) on a Pt electrode, fibrous PANi was obtained in 0.5 mol/L aniline, 0.5 mol/L $HClO_4$ aqueous solution. Granular PANi was produced when the $HClO_4$ electrolyte was replaced with H_2SO_4.[61] The morphology of PANi prepared in a 0.7 mol/L aniline, 1.5 mol/L HBF_4 aqueous solution by the constant current polymerization is also fibrous.[62,67] Fibrous PANi cannot be formed in an HCl or H_2SO_4 aqueous solution and in an HBF_4 acetonitrile solution.[62] The polymerization current can affect the PANi morphology: the diameter of the PANi fibers decreases with increasing polymerization current.[67] On ITO electrode by cyclic potential scanning between

−0.15 V and 0.78 V vs. SCE, fibrous and porous PANi was produced in acidic aqueous solutions of H_2SO_4, HCl, HNO_3, $HClO_4$, and CF_3CO_2H.[64,66] Mattosa et al.[65] found that the molecular weight of the PANi prepared electrochemically is lower than that synthesized by chemical polymerization. However, with the addition of neutral salt LiCl at a high concentration, the molecular weight of the electropolymerized PANi can be increased up to 160,000 g/mol.

Overoxidation is a major problem to deal with in the electyropolymerization of aniline in the acidic aqueous solutions.[63] At the polymerization potential of aniline, the polyaniline deposited on the electrode may be over-oxidized, which leads to partial dissolution of the film and the simultaneous formation of some insoluble residue. Park et al.[69] studied the electropolymerization of aniline by electrochemical quartz crystal microbalance (EQCM) and found that the current efficiency in a constant potential polymerization is lower than the cyclic potential scanning method, which indicates the existence of PANi overoxidation in the constant potential polymerization. Hence, cyclic potential scan in a potential range[64,69] (such as −0.15 V to 0.78 V vs. SCE[64]) is usually preferred for the preparation of a high quality PANi.

Usually, PANi shows electrochemical activity only in an acidic solution. It loses electrochemical activity in a neutral or basic solution, which is a serious problem for its applications such as in biosensors and as an electrode material. Mu prepared the copolymer of aniline and o-aminophenol by electrochemical polymerization in 0.2 mol/L aniline, 0.6 mol/L H_2SO_4 aqueous solution. The copolymer shows high electrochemical activity in aqueous solutions with pH value as high as 9.6.[70]

1.2.3.2 Electropolymerization in Nonaqueous Solutions

One promising application of the electrochemically prepared PANi is as the electrode materials in lithium or lithium ion batteries. Lithium batteries require a rigorous anhydrous environment. PANi prepared in acidic aqueous solutions always contains a certain amount of water. To overcome this problem, the electropolymerization of aniline in nonaqueous solutions was performed. Acidic media is still important for the electropolymerization of aniline in the nonaqueous solutions. In order to meet this requirement, anhydrous acid[71,72] or anilinium salt (formed by combining aniline and acidic compounds)[73,74] is usually utilized in the polymerization solutions. The solvents used include acetonitrile,[72,74] propylene carbonate (PC),[71,73] γ-butylolactone (GBL),[74] etc. The PANi prepared in the nonaqueous solution possesses high electroactivity when used as the electrode materials in Li/PANi batteries.[72,74] On the other hand, the conductivity of the PANi is on the order of 10^{-2} to 10^{-1} S/cm, lower than that deposited in the aqueous solutions.

Osaka et al.[71] prepared PANi in a CF_3COOH acidified PC solution. They found that the electroactive PANi films only formed in the solutions with the molar ratio of acid to aniline monomer greater than 1. With the increase of the acidity of the solutions, the resultant PANi film exhibits greater electroactivity and better charge-discharge performance of Li/PANi cathode cell. Pekmez et al.[72] carried out the electropolymerization of aniline in 0.1 mol/L aniline, 0.1 mol/L Bu_4NClO_4 acetonitrile solution by potential scanning between −0.2 V and 1.9 V vs. Ag/AgCl. They reported that adding anhydrous acid, such as $HClO_4$, to the solution improved the quality and yield of the PANi films. Furthermore, the concurrent generation of the acid through in situ oxidation of dissolved molecular hydrogen was the most efficient way of improving the polymer yield and properties. With saturated hydrogen in the solution, the conductivity of PANi was increased from 0.039 S/cm to 0.62 S/cm.[72] The electropolymerization of aniline can also be performed in organic solutions containing anilinium salt (such as aniline-HBF_4 salt[73,74]).

Pandey and Singh[75] prepared PANi in a proton-free nonaqueous solution. The polyaniline film made in dry acetonitrile containing sodium tetraphenylborate as supporting electrolyte is highly sensitive for proton sensing.

1.3 ELECTROCHEMICAL PROPERTIES OF CONDUCTING POLYMERS

Many applications of conducting polymers, such as electrode materials in batteries, modified electrodes, enzyme electrodes, electrochromics, electrocatalysis, and electrochemical sensors, are based on the electrochemical properties of the polymers. Furthermore, the HOMO (highest occupied molecular orbital) and LUMO (lowest unoccupied molecular orbital) of the optoelectronic conjugated polymers can be determined from the onset oxidation (p-doping) and onset reduction (n-doping) potentials of the conjugated polymers.[76] Therefore, electrochemical property is important in the studies and for the applications of conjugated polymers.

A unique characteristic of the electrochemical properties of conjugated polymers is that the p-doping/dedoping (oxidation/de-oxidation) reactions take place at a high potential range and the n-doping/dedoping (reduction/de-reduction) reactions occur at a low (negative) potential range. The conductivity of the polymers increases by 5 ~ 10 orders of magnitude upon doping (and hence the term "conducting polymers"). When the p-doping reaction takes place, the conjugated polymer chains are oxidized by removing electrons and counter-anions are doped into the oxidized chains to maintain electroneutrality:

$$CP - e^- + A^- \rightleftharpoons CP^+(A^-) \tag{1.15}$$

where CP represents the conjugated polymer and $CP^+(A^-)$ represents the oxidized conjugated polymer molecules doped with counter-anions of A^-.

When the n-doping reaction occurs, the conjugated polymer chains are reduced by injecting electrons to the chains. Counter-cations are doped into the reduced polymer chains to obtain electroneutrality:

$$CP + e^- + M^+ \rightleftharpoons CP^-(M^+) \tag{1.16}$$

Electrochemical methods used in the studies of conjugated polymers include cyclic voltammetry, in situ spectroelectrochemical method, EQCM, and electrochemical ac impedance spectroscopy. Cyclic voltammetry is usually used to study the reversibility of the redox reactions, measure the p-doping and n-doping potentials, and determine the HOMO and LUMO energy levels of the conjugated polymers. Electrochemical in situ UV-Vis absorption spectra can be used to understand the doping/de-doping states of the conjugated polymers during electrochemical redox processes,[77] which is crucial for elucidating the reaction mechanism of the redox processes of the conjugated polymers. EQCM measures the change of mass of the conjugated polymer films on the electrode during the electrochemical reaction. Since the doping/de-doping of conjugated polymers involves the uptake or ejection of counter-ions, the EQCM results can be directly correlated to the doping state of the polymers.[78] Electrochemical ac impedance spectra are useful in studying the characteristics of the electrode/electrolyte interface and the electrode kinetics including charge transfer rate and the diffusion coefficient of the counter-ions,[79] both of which play an essential role in the electrochemical applications of conducting polymers.

The electrochemical applications are mainly for the doped conducting polymers. PPy, PTh, and PANi are stable at the doped state. Therefore, we will mainly discuss the basic electrochemical properties of conducting PPy, PANi, and PTh. In fact, the electrochemical processes of conducting polymers are rather complicated. In addition to the charge transfer at the electrode/electrolyte interface and ionic diffusion in the electrolyte solution, the electrochemical processes also involve doping/de-doping and diffusion of the counter-ions in the conducting polymer films, conductivity change, volume expansion and shrinking of the polymer films, and the electron transfer between the polymer and the substrate electrode. Furthermore, different morphologies and structures of the polymer films prepared at different conditions also cause differences in their electrochemical responses.

1.3.1 Electrochemical Properties of PPy

PPy occupies an important position in conducting polymers, owing to its advantages of easy preparation of PPy films by electrochemical polymerization, environmental-friendliness, high stability at its conducting state, and high conductivity. The p-doping potential of PPy is low; that is, it is very easy to be p-doped. There is no report on its n-doping. Because the p-doped state is the most stable state of PPy, the electrochemical properties of PPy are mainly for the reduction (p-dedoping) and re-oxidation (p-doping) characteristics of the p-doped PPy. The reduction/re-oxidation of the p-doped PPy can be generalized as

$$PPy^+(A^-) + e^- \underset{p-doping}{\overset{p-dedoping}{\rightleftharpoons}} PPy^o + A^- \tag{1.17}$$

where $PPy^+(A^-)$ represents the p-doped PPy with A⁻ counter-anions and PPy^o represents the neutral conjugated PPy. The precise electrochemical processes of PPy are dependent on preparation conditions of the PPy films, nature of the counter-anions, and the type of the electrolyte solutions used for the electrochemical studies.

1.3.1.1   Electrochemical Properties of PPy in Aqueous Solutions

The reduction and re-oxidation processes and the reaction mechanism of PPy in aqueous solutions are influenced by the nature of counter-anions and pH value of the aqueous solutions.[77–83] In a weak acidic solution containing smaller anions of NO_3^-, Cl^-, ClO_4^-, the reduction and re-oxidation of PPy is reversible, and the reaction mechanism follows Equation (1.17). Figure 1.1b shows the cyclic voltammogram of $PPy(NO_3^-)$ in pH 3, 0.5 mol/L $NaNO_3$ aqueous solution. There are a couple of broad and reversible reduction/re-oxidation peaks in the potential range of 0.3 ~ 0.8 V vs. SCE. In a neutral aqueous solution, there are two reduction peaks in the potential range of 0.3 ~ 0.8 V vs. SCE, as shown in Figure 1.1a. The two reduction processes correspond to two doping structures in PPy.[77] In fact, there are also two reduction processes for PPy in an acidic solution, which overlap into a broad reduction peak. The in situ absorption spectra of $PPy(NO_3^-)$ in pH 3, 0.5 mol/L $NaNO_3$ aqueous solution clearly show the two reduction processes from 0.3 to –0.3 V and from –0.3 V to –0.8 V vs. SCE.[77]

If the counter-anions in PPy are small-sized anions and the solution anions are big-sized anions with long alkyl chains (such as TsO^-), the first reduction peak in the cyclic voltammograms is normal, but the re-oxidation process is irreversible due to the doping difficulty of the big-sized solution anions. If the counter-anions in PPy are big-sized anions such as TsO^-, the reduction potential of PPy will shift negatively due to the difficulty of dedoping. The belated reduction involves the doping of cations from the solution instead of the dedoping of the counter-anions:

$$PPy^+ (A^-) + e^- + M \rightleftharpoons PPy0(MA) \qquad (1.18)$$

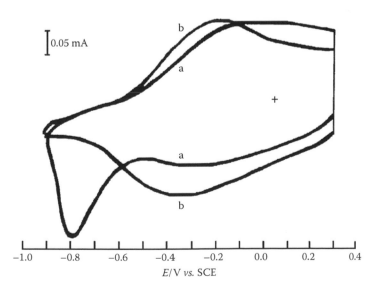

FIGURE 1.1   Cyclic voltammograms of $PPy(NO_3^-)$ in 0.5 mol/L $NaNO_3$ aqueous solutions with (a) pH 7 and (b) pH 3, scan rate = 20 mV/s.

In a basic solution, the counter-anions in PPy will exchange with the nucleophilic OH⁻ anions in the solution. The subsequent reduction/reoxidation involves the dedoping/doping of OH⁻.[83] In the meantime, the conjugated chain of PPy degrades during long duration of submersion in a basic solution. The degradation occurs rapidly in a strong basic solution, leading to the decrease or complete loss of conductivity and the electrochemical activity characteristic of PPy.

The upper potential limit in the cyclic voltammetry of PPy should be lower than 0.5 V vs. SCE. Otherwise, irreversible overoxidation of the p-doped PPy takes place, resulting in degradation and loss of electrochemical activity. The overoxidation reaction of PPy is affected by the solution pH value. The overoxidation potential decreases with increasing pH values,[84] that is, the overoxidation degradation of PPy is more readily to occur in a basic aqueous solution.

1.3.1.2   Electrochemical Properties of PPy in Organic Solutions

The cyclic voltammograms of PPy in an organic solution show interesting overpotential phenomenon in its first reduction,[85] which is different from that in aqueous solutions. For the cyclic voltammetry of the PPy films electrochemically prepared in an organic solution, the first reduction in an acetonitrile or propylene carbonate (PC) solution starts at approximately −0.6 V vs. SCE, which is much lower than the reduction potential of PPy in aqueous solutions. There is a strong reduction peak at approximately −0.8 V. The subsequent re-oxidation and the reduction/re-oxidation from the second circle in the cyclic voltammograms are the same with that in the aqueous solutions, as shown in Figure 1.2.[85] When the cyclic voltammetry of the PPy films prepared in organic solutions was performed in an aqueous solution, the overpotential phenomenon of the first reduction was not observed, indicating that the phenomenon is related to the salvation energy of the dedoped counter-anions of PPy into the electrolyte solutions. The explanation for the phenomenon is that the energy of the counter-anions dedoping into the organic solution is much higher than that dedoping into aqueous solutions. During the first reduction, the dedoping of

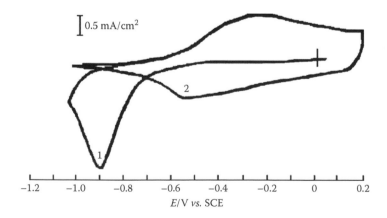

FIGURE 1.2   Cyclic voltammograms of PPy(ClO₄⁻) in 0.5 mol/L NaClO₄ PC solution. The numbers in the figure indicate the sequence of the potential scan.

the counter-anions into the organic solution is difficult, so a large negative potential is needed for the solvated cations in the solution to dope into the PPy film instead of the counter-anion dedoping. Then the doped cations dilate the PPy film, which increases the diffusion coefficient of the counter-anions. The cyclic voltammograms in the subsequent circles resume normal behavior (similar to those in aqueous solutions). The volume expansion of PPy films during the first reduction was observed in experimental measurements.[85]

1.3.2 ELECTROCHEMICAL PROPERTIES OF CONDUCTING POLYANILINE

The electrochemical properties of PANi are closely related to their preparation conditions and the pH values of the electrolyte solutions. Conducting PANi is dedoped and loses its electrochemical activity by deprotanation in neutral and basic aqueous solutions. Therefore, the electrochemical properties of PANi are usually studied in acidic aqueous solutions.

Figure 1.3 shows the cyclic voltammograms of PANi(NO_3^-) in pH 1.5, 1 mol/L NaNO$_3$ aqueous solution.[86] The PANi(NO_3^-) sample was prepared by electrochemical polymerization of aniline in HNO$_3$ aqueous solution at 1.0 V *vs.* SCE. It can be seen from Figure 1.3 that there are three pairs of redox peaks in the potential range of $-0.5 \sim 0.7$ V vs. SCE. According to the in situ absorption spectra at different potentials, the three pairs of redox peaks can be ascribed to the *p*-doping/dedoping of neutral PANi in the potential range of $-0.4 \sim 0.28$ V. Above 0.3 V, the doped PANi is further oxidized into bipolaron doping state with portions of PANi's

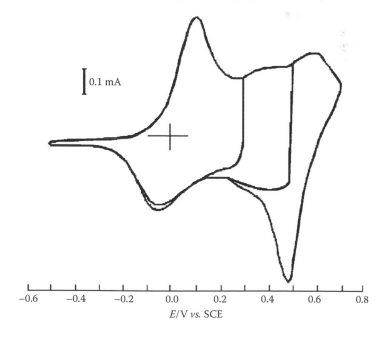

FIGURE 1.3 Cyclic voltammograms of PANi(NO_3^-) in pH 1.5, 1 mol/L NaNO$_3$ aqueous solution.

benzoid-amino structures oxidized into quinonoid-imino structures.[88] In comparison with PPy and polythiophene, the oxidation process in the potential range higher than 0.3 V is unique to PANi, due to the special benzoid-amino structure in PANi.

Genies et al.[87] studied the origin of the middle redox peaks of the three pairs of redox peaks of PANi, and found that the middle peaks can be accounted by the presence of an overoxidized PANi structure containing phenazine rings, which were formed during the polymerization of aniline at a rather high potential. At the high polymerization potential, greater than 0.8 V vs. SCE, PANi is probably partially overoxidized.

PANi also possesses electrochemical activity in organic electrolyte solutions. Watanabe et al.[88] studied the electrochemical polymerization of aniline in 0.1 mol/L $LiClO_4$ acetonitrile solution and the electrochemical properties of PANi in acetonitrile solution. There are two pairs of reversible redox peaks in the cyclic voltammograms of the PANi in the organic solution in the potential range of $-0.2 \sim 1.0$ V vs. SCE. The electrochemical activity in organic solutions is important for the application of PANi as an electrode material in lithium batteries.

The p-doped state of PANi is overoxidized at potentials higher than 0.7 V vs. SCE in an acidic aqueous solution.[86] Dinh et al.[89] also found that overoxidation degradation of PANi occurred at 1 V vs. NHE.

1.3.3   ELECTROCHEMICAL PROPERTIES OF POLYTHIOPHENE

PTh cannot only be oxidized (p-doped) in positive potential range but also can be reduced (n-doped) in negative potential range, owing to its narrower bandgap and higher oxidation (p-doping) potential, which is different from that of PPy and PANi. There is a pair of reversible oxidation/re-reduction (p-doping/dedoping) peaks in the cyclic voltammogram of PTh in acetonitrile solution in the potential range of $0.5 \sim 1.2$ V vs. SCE. The reaction mechanism of PTh should be very similar to that of PPy, but the p-doping potential of PTh is approximately 1 V, higher than that of PPy. The stability of the conducting p-doped PTh is lower than that of conducting PPy and PANi.[90] PANi has the highest p-doping potential and therefore its p-doped state is the most vulnerable to reduction or dedoping. On the other hand, the neutral PTh is semiconducting and fairly stable. PTh can be used as a semiconducting polymer in optoelectronic devices.

Soluble polythiophene derivatives have drawn great attention in recent years, due to their applications in polymer solar cells[91,92] and organic field effect transistors.[93] The substituents on the thiophene rings cannot only improve the solubility of the PTh derivatives, but also tune the electronic energy levels. The recent electrochemical studies are mainly performed on these PTh derivatives.[92,94]

Figure 1.4 shows the cyclic voltammograms of three PTh derivatives in 0.1 mol/L Bu_4NClO_4 acetonitrile solution.[94] There are a pair of reversible p-doping/dedoping peaks in the positive potential range, and a pair of reversible n-doping/dedoping peaks in the negative potential range. The substituents on the PTh derivatives can significantly shift the p-doping and n-doping potentials. In general, an electron-donating group (such as an alkoxyl group) shifts the p-doping potential to a more negative value, while an electron-withdrawing group shifts the potential in the positive direction.[95]

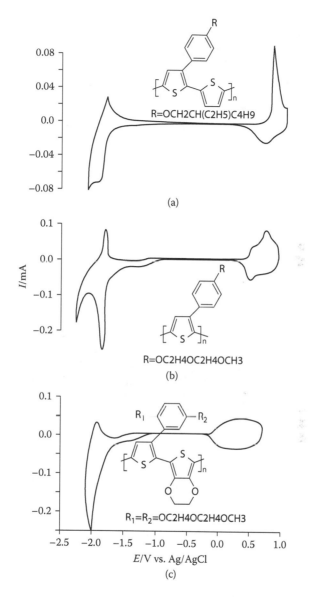

FIGURE 1.4 Cyclic voltammograms of polythiophene derivatives in 0.1 mol/L Bu_4NClO_4/ acetonitrile solution.

1.3.4 ELECTROCHEMICAL MEASUREMENT OF THE ELECTRONIC ENERGY LEVELS OF CONJUGATED POLYMERS

Electronic energy levels, including the LUMO and HOMO, of conjugated polymers are key parameters to consider for the applications of these polymers in optoelectronic devices such as polymer light-emitting diodes (PLEDs) and polymer solar cells (PSCs). An electrochemical method is a simple and convenient approach to measure the LUMO and HOMO energy levels.

The electrochemical measurement of the electronic energy levels is based on the assumptions that the ionization potential (IP) or the HOMO level of the polymers corresponds to their onset oxidation (p-doping) potentials (E_{ox}^{on}), and the electron affinity (EA) or the LUMO level corresponds to their onset reduction (n-doping) potentials (E_{red}^{on}).[76] The difference between the onset oxidation and reduction potentials is the bandgap (E_g) of the polymer. The onset oxidation and onset reduction potentials can be obtained from the cyclic voltammograms of the conjugated polymers. The HOMO and LUMO energy levels are calculated from the following equations:[96–98]

$$E_{HOMO} = \text{-IP} = -e(E_{ox}^{on} + C)(eV) \tag{1.19}$$

$$E_{LUMO} = \text{-EA} = -e(E_{red}^{on} + C)(eV) \tag{1.20}$$

$$Eg = e[E_{red}^{on} - E_{ox}^{on}] \; (eV) \tag{1.21}$$

where the unit of the potential values is volt. C is a constant, depending on the reference electrode (RE) used in measuring the onset oxidation and reduction potentials. C is equal to 4.8 when the RE is Fc/Fc$^+$ (ferrocene/ferrocium), 4.71 when RE is Ag/Ag$^+$, and 4.4 when RE is SCE.

1.4   SUMMARY

Electrochemistry is an important tool to synthesize and to characterize conjugated polymers. In the history of the development of conducting polymers, electropolymerization played a key role in the preparation of various conducting polymers, particularly PPy, PTh, and PANi. Unlike polyacetylene, these polymers are stable in the doped conductive state, and have been studied extensively for various applications as processable conductors or as electrodes in electrochemical and optoelectronic devices. The quality of the electropolymerized conducting polymer films is affected by several important reaction conditions in a complicated manner. Researchers may find electropolymerization being a science and an art at the same time. Conjugated polymers exhibit important electrochemical properties. The p-doping/dedoping and n-doping/dedoping are also affected by the measurement conditions and by the polymer preparation conditions. Conjugated polymers are vulnerable to overoxidation, during electropolymerization and during p-doping at a high potential. Overoxidation can be alleviated by careful control of the driving potential and through derivation of the monomers and polymers to reduce the electropolymerization and p-doping potentials.

REFERENCES

1. (a) Shirakawa, H.; Louis, E. L.; MacDiarmid, A. G.; Chiang, C. K.; Heeger, A. J. *J. Chem. Soc., Chem. Commun.*, 1977, 578. (b) Chiang, C. K.; Fincher, Jr. C. R.; Park, Y. W.; Heeger, A. J.; Shirakawa, H.; Louis, E. J.; Gau, S. C.; MacDiarmid, A. G.; *Phys. Rev. Lett.*, 1977, *39*, 1098
2. Tourillon, G.; and Garnier, F. *J. Electroanal. Chem.*, 1982, *135*, 173.
3. Dall'Olio, A.; Dasccola, Y.; Varcca, V.; Bocchi, V. *C. R. Acad. Sci., Ser. C*, 1968, *267*, 433.
4. Diaz, A. F.; Kanazawa, K. K.; Gardini, G. P. *J. Chem. Soc. Chem. Commun.*, 1979, 635.
5. Salmon, M.; Diaz, A. F.; Logan, A. J.; Krounbi, M.; Bargon, J. *Mol. Cryst. Liq. Cryst.*, 1982, *83*, 265.
6. Qian, R.Y.; Qiu, J. J. *Polym. J.*, 1987, *19*, 157.
7. Warren, L.F.; Anderson, D. P., *J. Electrochem. Soc.*, 1987, *134*, 101.
8. Satoh, M.; Kaneto, K.; Yoshino, K. *Synth. Met.*, 1986, *14*, 289.
9. Shen, Y. Q.; Qiu, J. J.; Qian, R. Y.; Carneiro, K., *Makromol. Chem.*, 1987, *188*, 2041.
10. Otero, T. F.; Santamaria, C. *Electrochim. Acta*, 1992, *37*, 297.
11. Zotti, G.; Schiavon, G.; Zecchin, S.; Sannicolo, F.; Brenna, E. *Chem. Mater.*, 1995, *7*, 1464.
12. Li, Y. F.; Yang, J. *J. Appl. Polym. Sci.*, 1997, *65*, 2739.
13. Ouyang, J. Y.; Li, Y. F. *Polymer*, 1997, *38*, 1971.
14. Wernet, W.; Monkenbusch, M.; Wegner, G. *Mol. Cryst. Liq. Cryst.*, 1995, *118*, 193.
15. Qian, R.Y.; Pei, Q. B.; Huang, Z. T. *Makromol. Chem.*, 1991, *192*, 1263.
16. Fukuyama, M.; Nanai, N.; Kojima, T.; Yasuo, Y.; Yoshmura, S. *Synth. Met.*, 1993, *58*, 367.
17. Kupila, E.-L.; Kankare, J. *Synth. Met.*, 1995, *74*, 241.
18. Ouyang, J. Y.; Li, Y. F. *Polymer*, 1997, *38*, 3997.
19. Li, Y. F.; Ouyang, J. Y. *Synth. Met.*, 2000, *113*, 23.
20. Li, Y. F.; Qian, R.Y. *Electrochim. Acta*, 2000, *45*, 1727.
21. Maddison, D.S.; Unsworth, J. *Synth. Met.*, 1989, *30*, 47.
22. Cvetko, B.F.; Brungs, M.P.; Burford, R. P.; Skallas-Kazacos, M. *J. Mater. Sci.*, 1988, *23*, 2102.
23. Liang, M.; Lei, J.; Martin, C. R. *Synth. Met.*, 1992, *52*, 227.
24. Li, Y. F.; He, G. F. *Synth. Met.*, 1998, *94*, 127.
25. Genies, E. M.; Bidan, G.; Diaz, A. F. *J. Electroanal. Chem.*, 1983, *149*, 101.
26. Li, Y. F. *J. Electroanal. Chem.*, 1997, *433*, 181.
27. Pei, Q.B.; Qian, R.Y. in *Polymers and Biomaterials*, edited by H. Feng, Y. Han, L. Huang, Elsevier Science Publishers, B.V., 1991, pp. 195–200.
28. Li, Y. F.; Fan, Y. F. *Synth. Met.*, 1996, *79*, 225.
29. Schiavon, G.; Zotti, G.; Comisso, N.; Berlin, A.; Pagani, G. *J. Phys. Chem.*, 1994, *98*, 4861.
30. Beck, F.; Oberst, M.; Jansen, R. *Electrochim. Acta*, 1990, *35*, 1841.
31. Waltman, R. J.; Bargon, J.; Diaz, A. F. *J. Phys. Chem.*, 1983, *87*, 1459.
32. Tourillon, G.; Garnier, F. *J. Electroanal. Chem.*, 1984, *161*, 51.
33. Tanaka, K.; Shichiri, T.; Wang, S.; Yamabe, T. *Synth. Met.*, 1988, *24*, 203.
34. Otero, T. F.; de Larreta-Azelain, E. *Polymer*, 1988, *29*, 1522.
35. Krische, B.; Zagorska, M. *Synth. Met.*, 1989, *33*, 257.
36. Imanishi, K.; Satoh, M.; Yasuda, Y.; Tsushima, R.; Aoki, S. *J. Electroanal. Chem.*, 1989, *260*, 469.
37. Ito, M.; Tsuruno, A.; Osawa, S.; Tanaka, K. *Polymer*, 1988, *29*, 1161.
38. Tanaka, K.; Shichiri, T.; Yamabe, T. *Synth. Met.*, 1986, *16*, 207.
39. Otero, T. F.; Rodriguez, J.; de Larreta-Azelain, E. *Polymer*, 1990, *31*, 220.
40. Osawa, S.; Ogawa, T.; Ito, M. *Synth. Met.*, 1997, *90*, 109.
41. Shi, G. Q.; Jin, S.; Xue, G.; Li, C. *Science* 1995, *276*, 994.
42. Li, X. H.; Li, Y. F. *J. Appl. Polym. Sci.*, 2003, *90*, 940.
43. Audebert, P.; Hapiot, P. *Synth. Met.*, 1995, *75*, 95.

44. Hillman, A. B.; Mallen, E. F. *J. Electroanal. Chem.*, 1987, *220*, 351.
45. Hu, X.; Wang, G. M.; Wong, T. K. S. *Synth. Met.*, 1999, *106*, 145.
46. Yumoto, Y.; Morishita, K.; Yoshimura, S. *Synth. Met.*, 1987, *18*, 203.
47. Bazzaoui, E. A.; Aeiyach, S.; Lacaze, P. C. *Synth. Met.*, 1996, *83*, 159.
48. Bock, A.; Topeters, A.; Kryschi, C. *Synth. Met.*, 1995, *75*, 133.
49. Inganas, O.; Liedberg, B.; Wu, C. *Synth. Met.*, 1985, *11*, 239.
50. Yumoto, Y.; Yoshimura, S. *Synth. Met.*, 1986, *13*, 185.
51. Glenis, S.; Benz, M.; LeGoff, E.; Kanatzidis, M. G.; DeGroot, D. C.; Schindler, J. L.; Kannewurf, C. R. *Synth. Met.*, 1995, *75*, 213.
52. Zotti, G.; Schiavon, G.; Berlin, A.; Pagani, G. *Synth. Met.*, 1993, *61*, 81.
53. Fall, M.; Aaron, J. J.; Sakmeche, N.; Dieng, M. M.; Jouini, M.; Aeiyach, S.; Lacroix, J. C.; Lacaze, P. C. *Synth. Met.*, 1998, *93*, 175.
54. Higgins, S. J.; Jones, C.; King, G.; Slack, K. H. D.; Petidy, S. *Synth. Met.*, 1996, *78*, 155.
55. Dogbeavou, R.; El-Mehdi, N.; Naudin, E.; Breau, L.; Belanger, D. *Synth. Met.*, 1997, *84*, 207.
56. Ballarin, B.; Seeber, R.; Tonelli, D.; Andreani, F.; Bizzarri, P. C.; Casa, C. D.; Salatelli, E. *Synth. Met.*, 1997, *88*, 7.
57. Zhou, L.; Xue, G. *Synth. Met.*, 1997, *87*, 193.
58. Xu, J.; Shi, G.; Xu, Z.; Chen, F.; Hong, Z. *J. Electroanal. Chem.*, 2001, *514*, 16.
59. Akoudad, S.; Roncali, J. *Synth. Met.*, 1998, *93*, 111.
60. Zotti, G.; Cattarin, S.; Comisso, N. *J. Electroanal. Chem.*, 1987, *235*, 259.
61. Taguchi, S.; Tanaka, K. *J. Power Sources*, 1987, *20*, 249.
62. Chen, S.-A.; Lee, T.-S. *J. Polym. Sci., C: Polym. Lett.*, 1987, *25*, 455.
63. Horanyi, G.; Inzelt, G. *J. Electroanal. Chem.*, 1989, *264*, 259.
64. Desilvestro, J.; Scheifele, W. *J. Mater. Chem.*, 1993, *3*, 263.
65. Mattosa, L. H. C.; Faria, R. M.; Bulhoes, L. O. S.; MacDiarmid, A. G. *Polymer*, 1994, *35*, 5104.
66. Bedekar, A. G.; Patil, S. F.; Patil, R. C.; Agashe, C. *Polym. J.*, 1995, *27*, 319.
67. Kamamura, K.; Kawai, Y.; Yonezawa, S.; Takehara, Z. *J. Electrochem. Soc.*, 1995, *142*, 2894.
68. Okamoto, H.; Okamoto, M.; Kotana, T. *Polymer*, 1998, *39*, 4359.
69. Choi, S.-J.; Park, S.-M. *J. Electrochem. Soc.*, 2002, *149*, E26.
70. Mu, S. L. *Synth. Met.*, 2004, *143*, 259.
71. Osaka, T.; Nakajima, T.; Naoi, K.; Owens, B. B. *J. Electrochem. Soc.*, 1990, *137*, 2139.
72. Pekmez, N.; Pekmez, K.; Arca, M.; Yildiz, A. *J. Electroanal. Chem.*, 1993, *353*, 237.
73. Takehara, Z.; Kanamura, K.; Yonezawa, S. *J. Electrochem. Soc.*, 1989, *136*, 2767.
74. Yonezawa, S.; Kanamura, K.; Takehara, Z. *J. Electrochem. Soc.*, 1995, *142*, 3309.
75. Pandey, P. C.; Singh, G. *J. Electrochem. Soc.*, 2002, *149*, D51.
76. Eckhardt, H.; Shacklette, L. W.; Jen, K. Y.; Elsenbaumer, R. L. *J. Chem. Phys.*, 1989, *91*, 1303.
77. Li, Y. F.; Qian, R. Y. *J. Electroanal. Chem.*, 1993, *362*, 267.
78. Li, Y. F.; Liu, Z. F. *Synth Met*, 1998, *94*, 131.
79. Li, Y. F.; Qian, R. Y. *Synth Met*, 1994, *64*, 241.
80. Kudoh. Y. *Synth. Met.*, 1996, *79*, 17.
81. Shimidzu, T.; Ohtani, A.; Iyoda, T.; Honda, K. *J. Electroanal. Chem.*, 1987, *224*, 123.
82. Li, Y. F.; Qian, R. Y. *Synth. Met.*, 1989, *28*, c127.
83. Li, Y. F.; Qian, R. Y. *Synth. Met.*, 1988, *26*, 139.
84. Li Y F, Qian R Y, *Electrochim Acta*, 2000, *45*, 1727–1731.
85. Li, Y. F. *Electrochim Acta*, 1997, *42*, 203.
86. Li, Y. F.; Yan, B. Z.; Yang, J.; Cao, Y.; Qian, R. Y., *Synth Met*, 1988, *25*, 79.
87. Genies, E. M.; Lapkowski, M.; Penneau, J. F. *J. Electroanal. Chem.*, 1988, *249*, 97.

88. Watanabe, A.; Mori, K.; Mikuni, M.; Nakamura, Y.; Matsuda, M. *Macromolecules*, 1989, *22*, 3323.
89. Dinh, H. N.; Ding, J. F.; Xia, S. J.; Birss, V. I. *J. Electroanal. Chem.*, 1998, *459*, 45.
90. Li, Y. F.; Qian, R. Y. *Synth Met*, 1993, *53*, 149.
91. Li, G.; Shrotriya, V.; Huang, J. S.; Yao, Y.; Moriarty, T.; Emery, K.; Yang, Y. *Nature Mater.* 2005, *4*, 864.
92. Hou, J. H.; Tan, Z. A.; Yan, Y.; He, Y. J.; Yang, C. H.; Li, Y. F. *J. Am. Chem. Soc.* 2006, *128*, 4911.
93. Ong, B. S.; Wu, Y.; Liu, P.; Gardner, S. *J. Am. Chem. Soc.* 2004, 126, 3378.
94. Johansson, T.; Mammo, W.; Svensson, M.; Andersson, M. R.; Inganas, O. *J. Mater. Chem.*, 2003, *13*, 1316.
95. Demanze, F.; Yassar, A.; Garnier, F. *Macomolecules*, 1996, *29*, 4267.
96. Prommerehne, J.; Vestweber, H.; Guss, W.; Mahrt, R. F.; Bassler, H.; Porsch, M.; Daub, J. *Adv. Mater.*, 1995, *7*, 551.
97. Li, Y. F.; Cao, Y.; Gao, J.; Wang, D. L.; Yu, G.; Heeger, A. J. *Synth. Met.*, 1999, *99*, 243.
98. Sun, Q. J.; Wang, H. Q.; Yang, C. H.; Li, Y. F. *J. Mater. Chem.*, 2003, *13*, 800.

2 From Metal to Semiconductor and Back—Thirty Years of Conjugated Polymer Electrochemistry

Olle Inganäs

CONTENTS

The transition from the semiconducting state to the metallic state—the SM transition—of conjugated polymers (CPs) under conditions of increasing doping is a perennial problem, part of the history of the research area of conjugated polymers. It is still perplexing that the semiconducting state of disordered polymers, now dominantly described as a state where hopping of charges between localized states occurs, to the metallic state of highly doped polymers, with increasing disorder, could at all be obtained in the same material. It would be even more surprising if this could be described in a single intellectual framework. The electrochemical functions of conjugated polymers rely on electronic charge transport and ionic charge transport—redox processes with all the concomitant processes due to co-transport of solvent, expansion/contraction of polymer chains, change of polarizability and surface energy, and first and foremost, change of electronic and ionic transport parameters. The change of electronic structure, electronic transport properties and

29

optical properties, ion transport conditions, and structural reorganization lead to a completely different material. The research traditions addressing the semiconductor-metal transition in highly doped CPs has many strands, and this account will be a rather personal and possibly eccentric selection of topics. A common theme is the use of electrochemical methods and concepts to advance the understanding of the SM transition. A desire is that of unifying the different modes of understanding electronic and ionic transport in solid-state electronics and in the electrochemistry of conjugated polymers.

2.1   INTELLECTUAL TRANSITIONS IN THE STUDY OF CP ELECTROCHEMISTRY

The electrochemistry of the newly discovered CPs of the early 1980s was peculiar in the combination of redox processes and capacitive charging of electrodes, combined in a complex pattern. Studies of Diaz[1–3] and many others focused on synthesis and the transformation of the insulating/semiconducting form of polymers such as poly-pyrrole, much in vogue at the time, due to the ease of preparing the polymer by electropolymerization. Electrochemical studies were typically done with liquid electrolytes, and a topic of fundamental interest was that of explaining the peculiar shape of cyclic voltammograms of these materials.[4] The conjugated polymers, which undergo redox transformation under insertion or removal of ions, can be considered as semiconductors or insulators in the pristine state. It was noted that at the fully doped state, capacitive behavior was dominating[4] and the electrode behaved as a porous metal electrode with non-Faradaic double layer charging. At negligible doping levels, the electrochemistry included Faradaic charge transfer causing a major change of phase of the materials. To reconcile these observations, the transition from semiconductor/insulator behavior to a metallic state was discussed. Somewhat later, introduction of well-defined oligomers of CP motifs, mainly phenylene vinylenes, enabled better-defined electrochemical studies,[5–8] and microelectrode studies of oligomers.[9] Here it was possible to unify the molecular electrochemistry point of view, where small oligomers allow one, two, or several more oxidation or reduction steps, depending on the length of the oligomer. With increasing oligomer length, the potential for the first electron transfer decreases in a systematic manner, and these oligomer properties could be used to correlate with the behavior of the polymer, which has a long conjugation length. As the oligomers grow in the length, they become less soluble, and technical complications limit the possibility of studying the electrochemistry of the long polymer chain in solution. These soluble polymer chains were, however, the objects that made the development of polymer electronics possible.

2.2   ELECTRONIC TRANSPORT AT DIFFERENT DOPING LEVELS

The semiconducting state is the state that has been the focus of attention to CPs since the discovery/assembly of semiconducting devices from conjugated polymers. These studies were products of parallel efforts to enable studies of the charge transport in CPs and to make semiconducting devices. Field effect transistors[10–12] became a growing field, which now has accumulated a considerable literature. Diodes[13,14] led to the discovery of electroluminescence in CPs,[15-17] also a field of industrial research at present. With the advancement of organic materials, diodes also came to include photodiodes and solar cells.[18-20] All three classes of devices mimic devices made from inorganic semiconductors, and the properties of inorganic devices are the benchmark by which development efforts are measured. The goals of this device development are those of suppressing instabilities and hysteresis, modifying interfaces for charge injection and extraction, engineering the sequence of layers for taming the injection, transport and recombination of charges, to feed into light emitting events, or conversely, to extract the photogenerated charges in photodiodes and solar cells. Removal of impurities, ions, solvents, and ambient gas are all of importance to optimize electronic functions, and thus to suppress the electrochemical functions of these materials.

The advancing understanding of electronic transport in the semiconducting CPs is evident in the development of transport models for these materials. While the focus on metallic transport in the 1980s made band models rather popular and proposed,[21] it became increasingly evident that these were not adequate to describe the mobility of charges in the semiconducting state. Extensive experimental studies of charge transport in semiconducting CPs, using diodes where bulk transport was limiting,[22,23] and in field effect transistors identified a mobility that was nonlinear in electric field and also concentration dependent. Models deriving from Heinz Bässler's studies of photophysics in disordered organics were adapted to charge transport in disordered solids,[24] and Monte Carlo modeling contributed to explain both field and concentration dependence of the mobility. These models were based on the assumption of complete localization of wave functions of electrons at sites, between which charge transport occurs by hopping (thermally assisted tunneling) between the localized states. These models were also used earlier in modeling of disordered semiconductors, and a considerable literature on electron hopping was already available and easy to tap into. It may also be noted that the focus on localized states was consistent with earlier strands of theory, in the form of the Anderson localization[25] of wave functions in one-dimensional systems, where any type of positional or energetic disorder leads to complete localization of the wave function. The dimensionality of the system has considerable implications for the localization properties, and two-dimensional interactions in combination with disorder may still allow some degree of delocalization, a topic that has received theoretical attention.[26]

The theoretical debate of the influence of order/disorder in CPs should also be seen in the context of the rapidly advancing materials science of CPs. The balance between crystalline order and poor processability and amorphousness with concomitant possibilities for low temperature processing took many steps between early studies of polyacetylene[27] and the present day plethora of processible and partially crystalline semiconducting polymers. With more and more of classical polymer physics applied and applicable to CPs, ordering and concomitant delocalization of wave functions over the ordered domain could be expected. This became experimentally visible through the two-dimensional delocalization of charge carriers on highly ordered poly(3-alkylthiophenes) in the solid state.[28]

With sufficient material order to allow delocalization on smaller domains, what would be the impact on the structure in these domains if charges are removed or added? First, on the single chain is the presence of polarons or bipolarons, a coupled charge and geometrical distortion of the dimerization pattern of the CPs. In electrochemical processes, with a source of ions nearby to the domain, influx or outflux of ions to compensate the electronic charge accommodates the redox transformation. Subsequent structural changes include the reorganization of ions in space, the flux of solvent (if a liquid electrolyte is used), and swelling/contraction of the polymer matrix.[29-31] None of these processes could be expected to enhance order in the CPs, but most will introduce further disorder. Conditions for localization therefore are enhanced upon electrochemical transformations. The semi-reversible transition from semiconductor to highly doped polymer therefore is associated with hysteresis and accumulation of disorder.

Extensive studies following the change of electronic structure and transport with doping levels are available.[32,33] Some of these studies have elaborated on the use of electrochemical transistors to delineate the density of states of the electronic conductor.[34,35] This is motivated by the efforts to unify the description of charge transport under varying density of charge carriers/doping. We note that the major structural changes under electrochemical transformation are a good reason to be skeptical of this ambition. Though it may be possible to overlap the density of states (DOS) as evaluated from field effect transistors and diodes with that of electrochemical transistors,[34] these are very different systems and should not be identified.

2.3   THE MOVING FRONT

In the transition from a doped to an undoped CP, or from an undoped to a doped CP, changes of electronic structure, electronic and ionic transport parameters, optical absorption, volume,[29,30] and chemical composition are present. The change most visible to the human eye is the change of color in absorption; with the combination of photoluminescence (PL) in the CPs, suppression of PL by doping gives a more sensitive visible expression of the impact of doping. The change of color was used for studying the transition by Aoki[36-42] in a series of elegant experiments in the late 1980s. At that time, most CPs were only available by chemical or electrochemical synthesis, and much of the experimental work on the doping front was performed using thin or thick CP electrodes ($\approx$0.1–10 µm), with the doping front moving in parallel to the electrode contact. The standard geometry was a thin film on an electrode, which allows for one-dimensional models. In the experiments of Aoki, the geometry used was a lateral one, with the doping front moving several millimeters over minutes and hours along a supporting substrate. The combined optical and electrochemical experiments required imaging of the propagating front of doping (or undoping) on a CP film supported on an insulating substrate immersed into a liquid electrolyte, and contacted at one end by an electronic conductor (Figure 2.1). A sharp boundary between doped and undoped regions could be identified, and the motion of this boundary under different driving conditions could be studied. A model of the propagation of the front based on electrochemical concepts was proposed and recently reasserted to explain some of the most anomalous (from the electrochemical point of

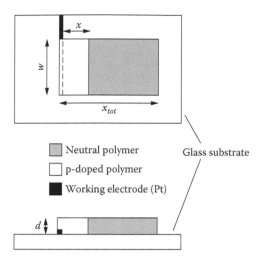

FIGURE 2.1 Outline of a moving front experiment. The polymer film is very thin (<100 nm) compared to the length and width. (From T. Johansson, N. K. Persson, and O. Inganas, *Journal of the Electrochemical Society*, 151, E119–E124, 2004. With permission.)

view) properties. We borrowed this method in our studies of the propagation of the moving front and applied it to several CPs,[43] some of which were also soluble.[44] This has the advantage that chemical synthesis of a soluble CP allows much better control over the chemical structure than electrochemical synthesis of ill-defined polymers on an electrode.

In these studies, a polymer film on an insulating substrate in contact with a liquid electrolyte, and with a metal contact at the edge of the film, was imaged with a microscope during the redox. In a three-electrode system, this process can be induced under potentiostatic control, and both current flow through the electrolyte and movement of a boundary between doped (blue) and undoped (red) polymer can be recorded as a function of time[44] for various applied doping potentials. The speed of the front can be evaluated, and correlation to current shows that the speed and current are linearly related. The time dependence of the front speed is proportional to $t^{-\gamma}$, where γ is 0.35 to 0.45, and varying with overpotential. At low overpotentials, no distinct boundary could be observed, and data was not recorded. In addition, during an experiment the boundary would become less and less distinct, until no data could be extracted. The study was therefore limited to rather high overpotentials.

Two aspects were emphasized in the analysis of the data. Clearly, the time dependence of the phase transformation and the resulting current had no similarity to simple models of diffusion in electrochemical systems; therefore, the electric field in the polymer layer must be accounted for. Second, the rate of conversion at the interface doped polymer/undoped polymer was clearly activated with increasing overpotential. A model incorporating the voltage drop along the polymer film, assumed to be constant within the doped section, was necessary to ascertain the real overpotential at the moving front position. The current generated by the movement of the front was found to be well described by a Tafel plot, and an anodic charge transfer

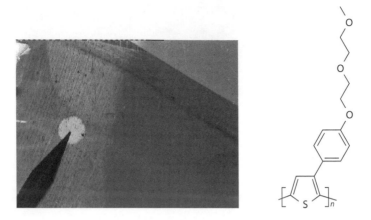

FIGURE 2.2   A two-dimensional moving front experiment; a thin film of the polythiophene derivative is oxidized from a point contact, and the boundary between the blue doped domain and the red undoped domain moves outward in concentric geometry. The film is immersed in a liquid electrolyte.

coefficient α from the Tafel plot was found to be 0.09, an anomalously low value in electrochemistry, but similar to α values derived from other polymers by Aoki. The anomalously low value was the subject of a paper from Aoki,[45] trying to show that this value originates from the asymmetry of the polymer matrix, being an electronic insulator and ionic conductor in one phase, and an electronic conductor in the other phase, using a Langevin equation.

An interesting alternative to the Aoki experiment is the variation of the same theme, devoted to studies of ion transport in CPs, as developed by Smela.[46–49] This geometry adds an ionically blocking layer on top of the CP, which is located on an electronic conductor. Electrochemical experiments are done under conditions where ion insertion can only occur from the edge of the barrier-covered film, in a two-dimensional geometry (Figure 2.2). These conditions can also create a front; it is essential to include drift/migration to model this in the continuum models for transport of electrons and ions that were presented. Interestingly, this model does not include a specific expression for ion/electron transfer at the electrode/electrolyte interface, but can still reproduce some of the phenomena reported from the electrochemical studies. However, these predictions require that migration is included in the model, arguing non-zero electrical field within the electrode.

2.4   LATERAL ELECTROCHEMISTRY IN DEVICES

The conversion of the undoped to the doped form of a CP can also be followed by the transition of conductivity from that of a very poor conductor to a highly conducting phase. Methods to follow this transition by the change of resistance in a thin polymer film have been used by many[50–52] and were introduced at an early stage through Wrighton's work on electrochemical transistors.[53–57] In these experiments, a thin film of CP is in contact with two solid electrodes, drain and source, and an electrolyte.

The electrolyte is contacted with a gate electrode, and a reference electrode may or may not be part of the setup. Applying a bias between the drain and source gives a current; that current can be modulated by applying a voltage versus the' gate, to dope or undope the polymer. The electrochemical transistor has become more fashionable in recent years[58,59] with the advancement of printed electronics on flexible substrates.[60–62] However, these devices all incorporate the semiconductor metal (SM) transition in their function, and therefore are vulnerable to our insufficient understanding of this transition. Models to predict the behavior of electrochemical transistors have been developed,[63,64] and device simulation modules have also been developed to match the needs of paper electronics. One of these models is based on a steady state solution of the Nernst equation at the electrode/electrolyte interface,[63] with continuum models of transport in adjacent electrode and electrolyte phases.

Light emitting electrochemical cells (LEC) were introduced by Pei et al.,[65] and are based on a combination of electrochemical and electronic processes in a CP. These devices often have a lateral geometry, with wide separation (10 to 10,000 μm!) between electrodes where charges are injected. They typically use a mixed electronic/ionic conductor, where the ionic conduction is enhanced by mixing polyethylene oxide or related compounds into a luminescent CP. Upon driving charges between the lateral electrodes on an insulating substrate, doping of the CP occurs, forming a front of p-doped materials propagating from the anode, and a front of n-type materials propagating from the cathode. Under lateral geometries and UV illumination, this can be observed easily,[66–68] as the doped regions change color and lose photoluminescence. Where p-front and n-front meet, light emission occurs due to charge recombination in a thin layer, somewhere between the anode and cathode. Clearly, the LEC contributes an experiment with moving n- and p-fronts simultaneously, and simple models of this experiment have been proposed.[67]

The mechanism at operation in these devices has been debated, as it has been suggested that the main effect of the electrolyte is that of modifying barriers for charge injection at electrodes.[69] A recent in situ Kelvin probe experiment has made imaging of the potential profile and imaging of the emission profile possible; this unequivocally shows the potential drop to occur far away from electrodes, somewhere in between anode and cathode, where a p-n junction is formed in a CP material.[70] This is a dynamic p-n junction, which can be changed by changing external voltage bias.

2.5 MODELING OF THE SM TRANSFORMATION INCLUDING ELECTROCHEMICAL ELEMENTS

Modeling of the moving front phenomena has been done by several scientists, including Lacroix, Fraoua, and Lacaze,[71] Robinson,[63] and very recently by Smela.[47] The continuum transport model proposed by Lacroix, Fraoua, and Lacaze[71] allows mobilities of ions and electrons to be widely different.

Modeling of the moving front phenomenon has been done with Monte Carlo simulations by Zuppiroli et al.[72,73] Here the change of electronic occupation at sites in a square lattice is accompanied by ion motion. A necessary condition for charge injection is the availability of a compensating ionic carrier at the site, and injection occurs from oxidized sites connected to an electrode. In an early model,[72] ions were assumed to transport by diffusion only (effectively claiming that the internal electric field is low). A later model included migration of ions in the time and space varying electric field inside the polymer layer.[73] Both of the models could reproduce features observed in experimental studies; the inclusion of migration elements also allowed description of situations where the formation of a space charge could prevent influx of the single charge ion, making complex behavior feasible.

These models may be attractive from a theorist's point of view, but they lack a representation of so many of the very real phenomena that are part of the SM transition. They do not include transport of solvent; they do not allow for the fact that the representation of transport processes on a lattice suppresses the fundamental condition that some of these lattice dots are connected by covalent bonds while others are not, that is, that the substrate matter is polymeric. Phenomena like the memory of electrochemical history in polymer electrodes, highlighted in studies by Nechtstein and coworkers,[74,75] and visible in slow relaxation phenomena which go beyond the time scale of microscopic processes, raise problems. The osmotic relaxation[31] occurring upon ion transport is not accounted for. Focusing on the coupling between mechanical degrees of freedom and the doping processes, for decades Otero has sought models incorporating elements of statistical physics and polymer physics with the electrochemistry of CPs.[76–78] These models are not microscopic, but include representations of geometry change and relaxation during redox of polymers. Without representation of these phenomena, a full account is presumably not possible, but the simplified analytical or numerical modeling available in simpler models still has a value in defining the limits to our understanding.

2.6   UNIFICATION OF ELECTRONICS AND ELECTROCHEMISTRY

Can we expect to see a unification of the theories of electronic and ionic transport in conjugated polymers, as seen from the electronic versus the electrochemical point of view? What is the real difference between the redox transformation of a single site on a polymer chain by oxidation, and the injection of a hole into the semiconducting polymer slab? As localization of the electronic wave function is (mostly) the case in disordered polymer solids, these are alternative descriptions of the same phenomena. What makes a real difference between electronic and electrochemical perspectives and systems is the availability of ions and solvents to diffuse/drift in the concentration gradients/fields associated with the charge injection. In semiconductor devices, these are processes giving rise to hysteresis, variation of device properties, and sources of instability, and many efforts are done to remove or minimize their influence. In electrochemistry, these modes are used. Intermediate hybrids, which utilize an electrolyte as gate contact to an organic semiconductor in an organic field effect transistor, have been presented.[79–81] These have the advantage of reducing the voltage necessary for enhancing a channel conductance, due to the high capacitance of the electrolyte/semiconductor contact; unfortunately, they also introduce the possibility of electrochemical doping. The introduction of the electrolyte gated field effect transistor[79–82] creates an opportunity to use both field effect doping and electrochemical doping. Anomalously high mobilities can be deduced from observations in such field effect transistors,[80,83] but a deeper analysis suggests that the standard models of mobility in field effect transistors is not simply applicable to such devices.[84]

As ion and small molecule diffusion are slow processes compared to electron tunneling and electronic transport, and also compared to the dynamics of the coupled electronic and vibration degrees of freedom of a CP, they might be associated with different time domains. We have observed this separation of electronic and ionic phenomena in an organic thin film transistor, where short times allow observation of a field effect, and where at longer times electrochemical doping of the organic cause a much larger current modulation. This transistor was based on a thin film coating a narrow insulating wire, in a fiber transistor. Here the crossing of wires, with an electrolyte separating the wires, forms a transistor. These devices are surprisingly resilient toward geometrical variations. As the insulator in a field effect transistor (FET) is substituted with an electrolyte, where the double layer capacitance very close to the semiconductor helps create the field doping, it does not matter much what the geometry of the electrolyte is. It can be thick and uneven; the potential drop always occurs close to the semiconductor/electrolyte interface.

Such devices are easy to prepare and make woven electronic structures possible. In one construction, the mode is mainly electrochemical and operates as a depletion mode transistor.[85] With a different choice of materials, enhancement mode transistors can be created.[86] Here we can use time resolved measurements on the drain source current to identify an early regime with a field effect modulated conductance. At later times (ms) there is the onset of electrochemical doping, giving much higher conductance.

The approximate model for the coupled electronic and ionic phenomena operating in all these different modes could be the following. At the metal/polymer interface, charge injection occurs to a site prepared for creation of localized charge. The injection is a tunneling or hopping event; the receiving state is a polaron, accessible by thermal fluctuation of the polymer chain geometry, and rapidly locked into the polaron geometry upon injection of charge. These effects are housed at the very first nanometer in the metal/polymer junction. Should ions be available to compensate for the injected charge, subsequent diffusion of ions (or solvent dipole dressed ions) will quench the space charge formed by this injection. Note that this is the injection process observed in polymer diodes, and typically used to determine the mobility of charge carriers, as the buildup of the space charge acts to prevent further injection, leading to the Childs space charge law ($J \approx V^2$). If ions are available, buildup of the space charge will be counteracted by the influx of compensating ions. This process will have a time response appropriate to the diffusion coefficient of ions in the solid, and will therefore be rather slow. In mixed ionic/electronic conductors, with a separate electrolyte phase, the fast diffusion of the ion in the electrolyte subsequently will be followed by slower diffusion in the electronic conductor of lower ionic conductivity; this then defines the appropriate length of diffusion necessary to accomplish space charge compensation. This charge compensation by ions stabilizes the polaron and further injection may occur. In addition, the domain housing a growing density of polarons grows out from the metal electrode, eventually forming a highly doped domain of high conductivity. The voltage drops precipitously as the electronic conductivity jumps many orders of magnitude, and the applied voltage is now active at the moving boundary of doped/undoped material. The rate of injection into polaron fluctuation states in the undoped polymer from the almost metallic polymer sets the stage for further motion. It has been shown that this motion can be described by a Tafel activation relation, but with an anomalously low charge transfer coefficient.[44,45]

Note that the rate of injection into the semiconductor polymer at the metal/polymer interface has been modeled in the context of electronic devices[87,88] and that at least one of these models connects the mobility in the bulk solid to the injection rate. A standard problem of electronic devices, where charge injection may be the limiting event at low voltages, to be superseded by space charge limitations in the bulk conduction at higher forward bias, is the coupling of these phenomena. For the electrolyte contact, we expect that, depending on the rates of electronic transport and ionic transport in the electrode compared to the rates of ionic transport in the electrolyte, we may come into very different situations of processes limited by the electronic transport in the electrode, or the ionic transport in the electrode or electrolyte. In our view, the fortuitous agreement of Tafel plots with the kinetics of the moving front, under some driving conditions, suggests that at high overpotential, the determining event is found at the interface between doped polymer/undoped polymer. If ionic supply is sufficient, the electronic transport is limiting in the semiconducting phase, and the injection of charge is ruled by simple exponential voltage dependence, which could be a Tafel expression or an injection expression.

2.7   CONCLUSION

In summary, a full integration of all the aspects of the semiconductor-metal transition in conjugated polymers has yet to be accomplished. Piecewise integration of the electronic and electrochemical interpretation of molecular domain charge transfer is on the way; full integration with aspects of conformation relaxation, however, is not completed. More lessons will be learned in the future.

2.8   ACKNOWLEDGMENTS

Dr. N.D. Robinson, IFM, Linköping University, helped with critical reading of this text. Research at the group for biomolecular and organic electronics is funded by the Science Council (VR), the Strategic Research Foundation SSF, and the Swedish Energy Agency (Energimyndigheten).

REFERENCES

1. A. F. Diaz, J. I. Castillo, J. A. Logan, and W. Y. Lee, *Journal of Electroanalytical Chemistry*, 1981, 129, 115–132.
2. A. F. Diaz, J. Castillo, K. K. Kanazawa, J. A. Logan, M. Salmon, and O. Fajardo, *Journal of Electroanalytical Chemistry*, 1982, 133, 233–239.
3. R. J. Waltman, J. Bargon, and A. F. Diaz, *Journal of Physical Chemistry*, 1983, 87, 1459–1463.
4. S. W. Feldberg, *Journal of the American Chemical Society*, 1984, 106, 4671–4674.
5. J. Heinze, J. Mortensen, K. Mullen, and R. Schenk, *Journal of the Chemical Society-Chemical Communications*, 1987, 701–703.

6. J. Heinze, M. Storzbach, and J. Mortensen, *Berichte Der Bunsen-Gesellschaft-Physical Chemistry Chemical Physics*, 1987, 91, 960–967.

7. K. Meerholz and J. Heinze, *Angewandte Chemie-International Edition in English*, 1990, 29, 692–695.

8. R. Schenk, H. Gregorius, K. Meerholz, J. Heinze, and K. Mullen, *Journal of the American Chemical Society*, 1991, 113, 2634–2647.

9. C. P. Andrieux, P. Audebert, P. Hapiot, and J. M. Saveant, *Journal of the American Chemical Society*, 1990, 112, 2439–2440.

10. A. Tsumura, H. Koezuka, and T. Ando, *Applied Physics Letters*, 1986, 49, 1210–1212.

11. A. Assadi, C. Svensson, M. Willander, and O. Inganas, *Applied Physics Letters*, 1988, 53, 195–197.

12. J. H. Burroughes, C. A. Jones, and R. H. Friend, *Nature*, 1988, 335, 137–141.

13. H. Tomozawa, D. Braun, S. Phillips, A. J. Heeger, and H. Kroemer, *Synthetic Metals*, 1987, 22, 63–69.

14. G. Gustafsson, M. Sundberg, O. Inganas, and C. Svensson, *Journal of Molecular Electronics*, 1990, 6, 105–111.

15. J. H. Burroughes, D. D. C. Bradley, A. R. Brown, R. N. Marks, K. Mackay, R. H. Friend, P. L. Burns, and A. B. Holmes, *Nature*, 1990, 347, 539–541.

16. D. Braun and A. J. Heeger, *Applied Physics Letters*, 1991, 58, 1982–1984.

17. R. H. Friend, R. W. Gymer, A. B. Holmes, J. H. Burroughes, R. N. Marks, C. Taliani, D. D. C. Bradley, D. A. Dos Santos, J. L. Bredas, M. Logdlund, and W. R. Salaneck, *Nature*, 1999, 397, 121–128.

18. N. S. Sariciftci, D. Braun, C. Zhang, V. I. Srdanov, A. J. Heeger, G. Stucky, and F. Wudl, *Applied Physics Letters*, 1993, 62, 585–587.

19. J. J. M. Halls, C. A. Walsh, N. C. Greenham, E. A. Marseglia, R. H. Friend, S. C. Moratti, and A. B. Holmes, *Nature*, 1995, 376, 498–500.

20. G. Yu and A. J. Heeger, *Journal of Applied Physics*, 1995, 78, 4510–4515.

21. A. J. Heeger, S. Kivelson, J. R. Schrieffer, and W. P. Su, *Reviews of Modern Physics*, 1988, 60, 781–850.

22. P. W. M. Blom, M. J. M. deJong, and J. J. M. Vleggaar, *Applied Physics Letters*, 1996, 68, 3308–3310.

23. P. W. M. Blom, M. J. M. deJong, and M. G. vanMunster, *Physical Review B*, 1997, 55, R656–R659.

24. H. Bassler, *Physica Status Solidi B-Basic Research*, 1993, 175, 15–56.

25. P. W. Anderson, *Physical Review* 1958, 109, 1492–1505.

26. M. Paulsson and S. Stafstrom, *Physical Review B*, 1999, 60, 7939–7943.

27. J. C. W. Chien, *Polyacetylene: Chemistry, Physics, and Material Science*, Academic, New York, 1984.

28. H. Sirringhaus, P. J. Brown, R. H. Friend, M. M. Nielsen, K. Bechgaard, B. M. W. Langeveld-Voss, A. J. H. Spiering, R. A. J. Janssen, E. W. Meijer, P. Herwig, and D. M. de Leeuw, *Nature*, 1999, 401, 685–688.

29. Q. B. Pei and O. Inganas, *Journal of Physical Chemistry*, 1992, 96, 10507–10514.

30. Q. B. Pei and O. Inganas, *Journal of Physical Chemistry*, 1993, 97, 6034–6041.

31. L. Bay, T. Jacobsen, S. Skaarup, and K. West, *Journal of Physical Chemistry B*, 2001, 105, 8492–8497.

32. X. Jiang, Y. Harima, K. Yamashita, Y. Tada, J. Ohshita, and A. Kunai, *Chemical Physics Letters*, 2002, 364, 616–620.

33. X. Q. Jiang, R. Patil, Y. Harima, J. Ohshita, and A. Kunai, *Journal of Physical Chemistry B*, 2005, 109, 221–229.

34. I. N. Hulea, H. B. Brom, A. J. Houtepen, D. Vanmaekelbergh, J. J. Kelly, and E. A. Meulenkamp, *Physical Review Letters*, 2004, 93.

35. I. N. Hulea, H. B. Brom, A. K. Mukherjee, and R. Menon, *Physical Review B*, 2005, 72.

36. Y. Tezuka and K. Aoki, *Journal of Electroanalytical Chemistry*, 1989, 273, 161–168.

37. K. Aoki, *Journal of Electroanalytical Chemistry*, 1990, 292, 53–62.

38. K. Aoki, *Journal of Electroanalytical Chemistry*, 1990, 292, 63–72.

39. K. Aoki, *Journal of Electroanalytical Chemistry*, 1991, 310, 1–12.

40. K. Aoki, *Journal of Electroanalytical Chemistry*, 1991, 300, 13–22.

41. Y. Tezuka, S. Ohyama, T. Ishii, and K. Aoki, *Bulletin of the Chemical Society of Japan*, 1991, 64, 2045–2051.

42. K. Aoki, T. Aramoto and Y. Hoshino, *Journal of Electroanalytical Chemistry*, 1992, 340, 127–135.

43. C. Carlberg, X. W. Chen, and O. Inganas, *Solid State Ionics*, 1996, 85, 73–78.

44. T. Johansson, N. K. Persson, and O. Inganas, *Journal of the Electrochemical Society*, 2004, 151, E119–E124.

45. K. Aoki, Journal of Electroanalytical Chemistry, 2004, 569, 121–125.

46. X. Z. Wang, B. Shapiro, and E. Smela, *Advanced Materials*, 2004, 16, 1605.

47. X. Z. Wang, B. Shapiro, and E. Smela, *Journal of Physical Chemistry C*, 2009, 113, 382–401.

48. X. Z. Wang and E. Smela, *Journal of Physical Chemistry C*, 2009, 113, 359–368.

49. X. Z. Wang and E. Smela, *Journal of Physical Chemistry C*, 2009, 113, 369–381.

50. G. Zotti, *Synthetic Metals*, 1998, 97, 267–272.

51. M. Lapkowski and A. Pron, *Synthetic Metals*, 2000, 110, 79–83.

52. T. Johansson, L. A. A. Pettersson, and O. Inganas, *Synthetic Metals*, 2002, 129, 269–274.

53. G. P. Kittlesen, H. S. White, and M. S. Wrighton, *Journal of the American Chemical Society*, 1984, 106, 7389–7396.

54. H. S. White, G. P. Kittlesen, and M. S. Wrighton, *Journal of the American Chemical Society*, 1984, 106, 5375–5377.

55. J. W. Thackeray, H. S. White, and M. S. Wrighton, *Journal of Physical Chemistry*, 1985, 89, 5133–5140.

56. E. W. Paul, A. J. Ricco, and M. S. Wrighton, *Journal of Physical Chemistry*, 1985, 89, 1441–1447.

57. S. Chao and M. S. Wrighton, *Journal of the American Chemical Society*, 1987, 109, 2197–2199.

58. D. Nilsson, M. X. Chen, T. Kugler, T. Remonen, M. Armgarth, and M. Berggren, *Advanced Materials*, 2002, 14, 51–54.

59. D. Nilsson, T. Kugler, P. O. Svensson, and M. Berggren, *Sensors and Actuators B-Chemical*, 2002, 86, 193–197.

60. P. Andersson, D. Nilsson, P. O. Svensson, M. X. Chen, A. Malmstrom, T. Remonen, T. Kugler, and M. Berggren, *Advanced Materials*, 2002, 14, 1460.

61. D. Nilsson, N. Robinson, M. Berggren, and R. Forchheimer, *Advanced Materials*, 2005, 17, 353.

62. P. Andersson, R. Forchheimer, P. Tehrani, and M. Berggren, *Advanced Functional Materials*, 2007, 17, 3074–3082.

63. N. D. Robinson, P. O. Svensson, D. Nilsson, and M. Berggren, *Journal of the Electrochemical Society*, 2006, 153, H39–H44.

64. D. A. Bernards and G. G. Malliaras, *Advanced Functional Materials*, 2007, 17, 3538–3544.

65. Q. B. Pei, G. Yu, C. Zhang, Y. Yang, and A. J. Heeger, *Science*, 1995, 269, 1086–1088.

66. D. J. Dick, A. J. Heeger, Y. Yang, and Q. B. Pei, *Advanced Materials*, 1996, 8, 985–987.

67. N. D. Robinson, J. H. Shin, M. Berggren, and L. Edman, *Physical Review B*, 2006, 74.

68. N. D. Robinson, J. F. Fang, P. Matyba, and L. Edman, *Physical Review B*, 2008, 78.
69. J. C. deMello, N. Tessler, S. C. Graham, and R. H. Friend, *Physical Review B*, 1998, 57, 12951–12963.
70. P. Matyba, K. Maturova, M. Kemerink, N. D. Robinson, and L. Edman, *Nat Mater*, 2009, advanced online publication.
71. J. C. Lacroix, K. Fraoua, and P. C. Lacaze, *Journal of Electroanalytical Chemistry*, 1998, 444, 83–93.
72. F. Miomandre, M. N. Bussac, E. Vieil, and L. Zuppiroli, *Electrochimica Acta*, 1999, 44, 2019–2024.
73. F. Miomandre, M. N. Bussac, E. Vieil, and L. Zuppiroli, *Chemical Physics*, 2000, 255, 291–300.
74. B. Villeret and M. Nechtschein, *Physical Review Letters*, 1989, 63, 1285–1287.
75. C. Odin and M. Nechtschein, *Physical Review Letters*, 1991, 67, 1114–1117.
76. T. F. Otero, H. Grande, and J. Rodriguez, *Journal of Electroanalytical Chemistry*, 1995, 394, 211–216.
77. T. F. Otero and E. Angulo, *Solid State Ionics*, 1993, 63–5, 803–809.
78. T. F. Otero and J. M. G. de Otazo, *Synthetic Metals*, 2009, 159, 681–688.
79. M. J. Panzer and C. D. Frisbie, *Journal of the American Chemical Society*, 2005, 127, 6960–6961.
80. A. S. Dhoot, G. M. Wang, D. Moses, and A. J. Heeger, *Physical Review Letters*, 2006, 96.
81. L. Herlogsson, X. Crispin, N. D. Robinson, M. Sandberg, O. J. Hagel, G. Gustafsson, and M. Berggren, *Advanced Materials*, 2007, 19, 97.
82. E. Said, X. Crispin, L. Herlogsson, S. Elhag, N. D. Robinson, and M. Berggren, *Applied Physics Letters*, 2006, 89.
83. A. S. Dhoot, J. D. Yuen, M. Heeney, I. McCulloch, D. Moses, and A. J. Heeger, *Proceedings of the National Academy of Sciences of the United States of America*, 2006, 103, 11834–11837.
84. J. D. Yuen, A. S. Dhoot, E. B. Namdas, N. E. Coates, M. Heeney, I. McCulloch, D. Moses, and A. J. Heeger, *Journal of the American Chemical Society*, 2007, 129, 14367–14371.
85. M. Hamedi, R. Forchheimer, and O. Inganas, *Nature Materials*, 2007, 6, 357–362.
86. M. Hamedi, L. Herlogsson, X. Crispin, R. Marcilla, M. Berggren, and O. Inganas, *Advanced Materials*, 2009, 21, 573.
87. J. C. Scott and G. G. Malliaras, *Chemical Physics Letters*, 1999, 299, 115–119.
88. Y. L. Shen, M. W. Klein, D. B. Jacobs, J. C. Scott, and G. G. Malliaras, *Physical Review Letters*, 2001, 86, 3867–3870.

3 Development and Applications of Ion-Functionalized Conjugated Polymers

David P. Stay, Stephen G. Robinson,
and Mark C. Lonergan

CONTENTS

3.1   INTRODUCTION

Ions are central to the chemistry and physics of conjugated polymers (CPs). The rapid escalation of interest in conjugated organic materials beginning in the late 1970s was initiated by an ion-containing material, namely doped polyacetylene.[1] The oxidation or reduction of polyacetylene was observed to result in an orders-of-magnitude increase in conductivity to levels rivaling those of metals. Such doping results in the incorporation of dopant counter ions that bear some analogy to ionized dopant centers in traditional inorganic semiconductors.[2] The dopant ions in CPs are typically peripheral to the extended bonding network rather than substitutional, as in the doping of inorganic semiconductors. There are many potential applications of CPs where ions play little or no role, such as in certain light-emitting diodes,[3] but ions are responsible for many of the unique properties, limitations, and potential of these materials relative to traditional inorganic semiconductors.

Ionic functionalization is particularly germane to ion-containing CPs because it affords a level of chemical control over the type and distribution of ions not possible with materials that are more conventional. Such control is important for understanding the role of ions in organic semiconductors and for applications based on ion-containing organic semiconductors. As with their non-conjugated counterparts, the introduction of ionic functionality has a substantial influence on solubility, can lead to properties strongly dependent on ion aggregation phenomena, imparts some level of ionic conductivity, and creates the potential for ion exchange processes. In addition, the electroactivity of CPs greatly expands the influence of ionic functional groups because these groups can compensate injected charge, shape the electrostatic potential distribution in materials under bias or at junctions, determine redox activity, and interact with photoexcitations.

Ion-functionalized CPs are most commonly referred to in the current literature as conjugated polyelectrolytes[4] or self-doped polymers,[5] although other terms such as ionomers have been used.[6–9] The literature on non-conjugated ion-functionalized polymers uses a wide range of terms to distinguish different classes based on the density and location of ionic functionality and material properties.[10] For instance, ionomers and polyelectrolytes are distinguished based on the relative importance of ion aggregation in determining physical properties,[11,12] which also leads to generalities regarding ion content as in the IUPAC definitions (ionomer as a low ion density material relative to a polyelectrolyte).[13] At present, there has been little effort to similarly classify ion-functionalized CPs. As the field advances, it will likely be beneficial to continue building connections to the conventional polyelectrolyte literature.[10] Throughout this review, the acronym IFCP (ion-functionalized conjugated polymer) is used to refer to any conjugated polymer with covalently bound ionic functional groups, regardless of the density of these groups or the physical properties they impart. IFCPs with covalently bound cationic and anionic functional groups are referred to as cationic and anionic IFCPs, respectively.

The backbone of a simple doped conjugated polymer is ionic and perhaps could be considered as belonging within a broad definition to an IFCP (or perhaps a conjugated ionene). For example, the polymer backbone of iodine-doped polyacetylene is charged with delocalized carbocations. Here, we restrict the definition of IFCPs to materials where the ionic functionality is peripheral rather than integral to the conjugated network of sp²-hybridized atoms characterizing CPs. In addition, we discuss IFCPs in the context that the ionizable groups are redox inactive, unlike doped sites. There is also work where CPs are used as ligands for redox active ions,[14–17] but this work is not covered.

There have been a number of excellent reviews covering IFCPs.[4,5,18–23] The goal of this review is to highlight the distinguishing features of IFCPs relative to non-ionic CPs. Consequently, it is organized around the physical phenomena that define these distinguishing features. In this way, we hope it complements other previously published reviews, which have been organized around the type of conjugated polymer backbone[5,18] or application.[4,19–23] Given space restrictions, only a subset of the available literature on IFCPs is discussed, again with an emphasis on highlighting the unique features imparted by ionic functionality. There are a number of materials whose solid-state phenomena are quite similar to IFCPs. Most notable are blends of CPs with non-conjugated polyelectrolytes[24–28] and small molecule, ionically functionalized organic semiconductors.[29] These materials are not covered, but some specific examples are given where illustrative. In addition, we do not cover ionically functionalized polyanilines because of the unique chemistry of polyaniline among CPs and because there has been an recent extensive review of their properties.[5]

3.2 SYNTHESIS

The general approaches used to synthesize IFCPs follow the same procedures as their neutral conjugated polymer cousins, which have been extensively reviewed.[3,30–34] The most common backbones for IFCPs are variations on poly(phenylene vinylene) (PPV),[35–38] poly(fluorene)(PF),[39–41] poly(thiophene)(PT),[42–44] and poly(para-phenylene) (PPP).[45–48] Less commonly reported are poly(pyrrole) (PP),[49–51] poly(carbazole),[52] poly(acetylene) (PA),[7] and poly(arylene-ethynylene) (PPE).[53] The most widely used ionic functional groups in IFCPs are sulfonate (SO_3^-),[35,42] carboxylate (CO_2^-),[44,45] and alkylammonium (NR_4^+),[41] although other functional groups such as phosphonates (PO_3^{2-}),[54,55] alkylpyridines,[47] and cationic phospholiums[56] are also found.

M = Na⁺ or H⁺

1

In the synthesis of IFCPs, ionic functional groups are introduced either before or after polymerization. Installation of ionic functionality before polymerization has advantages in that well-defined ion-containing monomers can be used. Such monomers give more control over the ionic functional groups in the final polymer than with the post-polymerization strategy where ensuring complete conversion can be difficult.[39] An advantage of post-polymerization functionalization is that the extensively studied polymerization conditions and characterization techniques for neutral

CPs can still be applied. For instance, Patil et al. found in their original syntheses of the family of anionic PTs **1** that not all of their ionic monomers (M = Na[+], n = 1) could be directly polymerized with standard polythiophene oxidative polymerization protocols.[42] This complication led them to use a post-polymerization functionalization strategy. Post-polymerization functionalization can also be advantageous in molecular weight determination as ionic functionality often leads to aggregation or specific interactions with the size-exclusion columns commonly employed for molecular weight determination.[7,39,43] The post-polymerization approach, however, does require the use of a monomer with a functional group that can be easily converted to an ionic functional group. Popular choices include esters[36] and phosponate esters,[54] which can be converted to carboxylates and phosphonates (often via their silyl esters) by hydrolysis; alkyl halides, which can be converted to, for instance, alkylammoniums by nucleophilic substitution;[47] or amines, which can be quaternized to their ammonium derivatives.[38]

3.3   IONIC FORCES AND INTERMOLECULAR INTERACTIONS

The most obvious consequence of ionic groups in CPs is the introduction of ion-dipole and ion-ion forces. These forces can result in strong interactions between polymer chains, different segments of the same polymer chain, and with other molecules.[10] Thus, ionic forces are being explored as a method to control the solution and solid-state properties of CPs and their interaction with other molecules, in particular for sensing applications and in the construction of assemblies. We present our discussion of ionic forces in four categories: solubility, solution conformation and aggregation, complex formation, and assembly. In many ways, these are all part of the same continuum, distinguished largely by context because a given counter ion could influence solution properties such as solubility or chain conformation in a relatively non-specific manner, have a more specific means of interacting with the IFCP upon binding to form a functional complex, or be considered part of a larger assembly of charged building blocks.

3.3.1   SOLUBILITY

Solubility distinguishes CPs from common inorganic semiconductors; consequently, it has been an important area of study. Indeed, water solubility was the motivation for one of the earliest reports on IFCPs by Patil et al., where they reported the synthesis of the anionically functionalized PT **1**.[42] Solubility continues to be a dominant theme in the development of IFCPs because of the impetus for low-cost solution processing in semiconductor technology.[57,58] Ionic functional groups drive solubility in polar solvents due to strong ion-dipole interactions and the potential for ion dissociation. These factors, despite the usually hydrophobic nature of the CP backbone, often result in IFCPs being soluble in water. This solubility is of particular importance in biological and environmental applications.

Nearly all of the undoped IFCPs that have been reported are soluble in either polar protic or aprotic solvents. In addition, a number of ionically functionalized PTs[42,59] and PPs[51,60,61] have been reported to be soluble in their doped forms. Upper limits on solubility are rarely reported, but reports of solution concentrations in the 1 to 50 mg/ mL range are common.[21,39,62–64] Synthetic control over the type and density of ionic functional groups provides a means for tuning the solubility and solution properties of IFCPs. In an example from our own work on PA derivatives, we have found that the solubility of sulfonate functionalized PAs can be tuned from dichloromethane, to dimethylformamide, to methanol, to water by changing the ionic functional group density and tuning the counter cation.[7,65,66]

The solubility characteristics of IFCPs have proven useful in a number of applications. For instance, solubility in polar solvents has been used in the fabrication of multilayer devices where the IFCP is partnered with non-ionically functionalized materials, which are typically soluble in non-polar solvents.[67,68] This partnering makes possible the sequential solution-based deposition of two different materials with minimal disruption of the first deposited layer due to orthogonal solubility. Specific examples can be found in Section 3.6.3 when discussing transport layers. The control of solubility through ionic functionality has also been important in efforts to sense aqueous analytes using changes in the spectroscopic properties of CPs. For instance, ionic functional groups have been used to solubilize CPs for aqueous-phase fluorescence quenching assays, such as in the sensing of hydrogen peroxide[69] or mercury ions.[70] Ionic functionality has also been used directly in sensing assays as a means to bind analyte or probe molecules as discussed in Section 3.3.3.

The amphiphilic nature of IFCPs can lead to complex solution behavior, which tends to limit exposure of the usually hydrophobic core or polar ionic functional groups to polar or non-polar solvents, respectively.[71,72] Aspects of this solution behavior are discussed in the next section. Concerning solubility, the amphiphilic nature of IFCPs has been used simply as an electroactive solubilizing agent for materials relevant to the development of organic electronic devices. For instance, Cheng et al. used both an anionic PPP (2) and an anionic PPE (3) to solubilize single-walled carbon nanotubes.[73] They showed that a stable carbon nanotube-IFCP complex could be obtained due to the strong supramolecular association between the carbon nanotube and conjugated polymer backbone. The ionic functional groups attached to the IFCP were then capable of solvating the structure in water, and excess, unassociated polymer could be removed by filtration.

Much as with more conventional polyelectrolytes,[74,75] IFCPs can also be hygroscopic.[76,77] Ortony et al. studied the hygroscopicity of a family of cationic PFs (4) with the same backbone in every case but varying the counter anion.[77] Comparison of the dehydration enthalpies showed the polymer with the smallest anion, 4 (X=Br), to be the most hygroscopic and that with the largest, 4 (X= B(ArF)$_4^-$), to be the least hygroscopic. The affinity of polyelectrolytes for water and other polar solvents can make it more difficult to remove solvents completely from solution cast films than with nonionic polymers.

3.3.2 SOLUTION CHAIN CONFORMATION AND AGGREGATION

Understanding chain conformation and aggregation of ion-functionalized polymers in solution remains a challenging topic because of the interplay between long-chain molecular structure and long-range ion-ion interactions.[10] The structure of ion-functionalized polymers in solution depends in a complex way on the nature of the polymer, including the density and type of ionic functional groups, solvent polarity, additives such as a background electrolyte, and polymer concentration. It is also common for chains to associate into larger structures.[10,78]

IFCPs are expected to exhibit much of the complex solution-phase behavior that typifies more common ion-functionalized polymers. The association of chains is of particular importance to IFCPs because interchain interactions contribute to the optical and electrical properties of CPs. In fact, fluorescence and absorption spectroscopy are commonly used to infer the formation of aggregates with strong electronic communication between conjugated segments and to probe conformational changes via associated changes in conjugation length. As they are not direct structural probes, it can be difficult to use these techniques to unambiguously differentiate changes in conformation vs. aggregation, as well as inter- vs. intrachain contributions to aggregate states; analyzing concentration dependence and pairing with other techniques, such as scattering methods,[10] can be helpful in this regard. In the following, we provide a few examples that illustrate the dependence of molecular structure on the density of ionic functional groups, the nature of the counter ion, and the effect of additives.

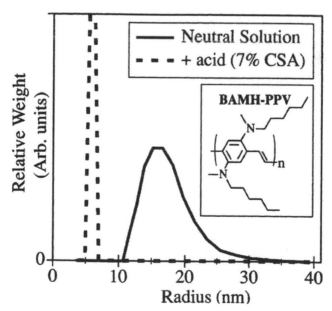

5

The introduction of even a small number of ionic functional groups can dramatically alter chain conformation and aggregation. Nguyen et al. studied changes in the conformation and aggregation of an amine functionalized PPV (**5**) upon protonation in o-xylene to form an IFCP.[8] In a dilute polymer solution, they observed the hydrodynamic radius distribution substantially narrow and shrink to approximately 6 nm upon protonation of approximately 10% of the amine sites (see Figure 3.1). The absorption and photoluminescence (PL) spectra of these solutions were also observed to blueshift by 50 nm or more and the PL quantum yield was observed to decrease from 79% to 35% upon protonation. These results were argued to be due to a change from a relatively open-chain conformation to a more tightly coiled conformation upon protonation to minimize exposure to

FIGURE 3.1 Dynamic light scattering size distributions (hydrodynamic radii) for dilute neutral (solid curve) and protonated (7% w/w/ camphor sulfonic acid, dashed curve) solutions of **5** in o-xylene. The inset shows the chemical structure of **5**. Reprinted with permission from [8]. Copyright 2002, American Institute of Physics.

the non-polar solvent. In solutions that are more concentrated, protonation was observed to result in a redshift in the PL spectrum relative to the neutral form, which was argued to be due to an increase in the aggregation state of the polymer. The formation of large aggregates was also implicated as an explanation for the much greater roughness observed for spin-cast thin films of the protonated relative to the neutral form.

PL, absorption spectroscopy, and light scattering have been used to probe the dependence of IFCP chain aggregation on solvent polarity, counter-ion, ionic strength, and solvent history.[9,76,79–81] Yang et al. reported on how the extent of solution aggregation can be tuned by counterion choice.[80] They used dynamic light scattering to show that with the cationic PF **6,** small anions such as Br led to greater aggregation while larger anions such as $B(Im)_4^-$ or $B(ArF)_4^-$ led to less aggregation, especially in water. Their explanation was that the large bulky anions act as spacers that prevent the close association of the polymer chains. The PL quantum yield in water varied from 3 to 22%, and the PL emission maximum varied from 589 to 572 nm in water, with the $B(ArF)_4^-$ yielding the largest quantum yield and the shortest wavelength PL emission maximum. Garcia and Nguyen observed two broad peaks in the absorption spectrum of the anionic PT **7** in water, and they assigned the longer wavelength peak to aggregates as supported by the fact that it became more prevalent as the polymer concentration increased.[81] They found that once formed, the aggregate peak could be difficult to disrupt with dilution, leading the absorption spectrum to depend not only on concentration, but also on the history of the solution. As seen in Figure 3.2, the aggregate peak could be diminished by the addition of excess salts with bulky counter-cations (tetrabutylammonium or benzyltriethylammonium) or polyethylene glycol; the latter was also effective in improving the quality of spin-cast films.

7

The addition of anionic,[82] cationic,[83] and non-ionic[84] surfactants has been shown to solubilize IFCPs and strongly influence their solution aggregation. For instance, Knaapila et al. showed that the addition of pentaethyleneglycol monododecyl ether to the anionic PF **8** solubilized the IFCP in water.[84] They further used PL, surface tension, static contact angle, π-A isotherm measurements, and small angle neutron scattering to study the nature of the solution aggregates formed upon solubilizing the polymer. They observed complex aggregate behavior with several different phases, depending on the concentration of surfactant.

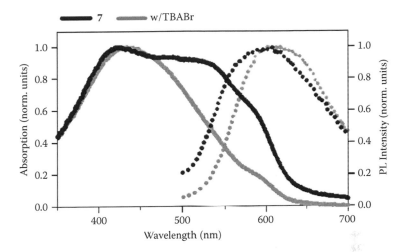

FIGURE 3.2 Absorption (solid lines) and photoluminescence (dotted lines) of **7** in water (black) and with addition of tetrabutylammonium bromide (grey). Data for an additional salt were omitted for clarity and can be found in the original reference. (Adapted from Chen, L. H.; Xu, S.; McBranch, D.; Whitten, D. 2000. Tuning the properties of conjugated polyelectrolytes through surfactant complexation. *J. Am. Chem. Soc. 122*: 9302–9303. With permission.)

The fluorescence changes of dissolved IFCPs induced by changes in solution state configuration can be used in sensing applications. For example, the use of self-quenching due to chain aggregation was reported by Satrijo et al. to sense small molecule multicationic biological molecules.[85] Specifically, anionic PPE **9** was used to detect analytes like spermine, spermidine, or neomycin by measuring the fluorescence changes that occurred from aggregation of the IFCP in the presence of these analytes. This assay relied on relatively non-specific interactions. Sensing assays that introduce a more specific means of analyte recognition through complexing to probe molecules are discussed in the next section.

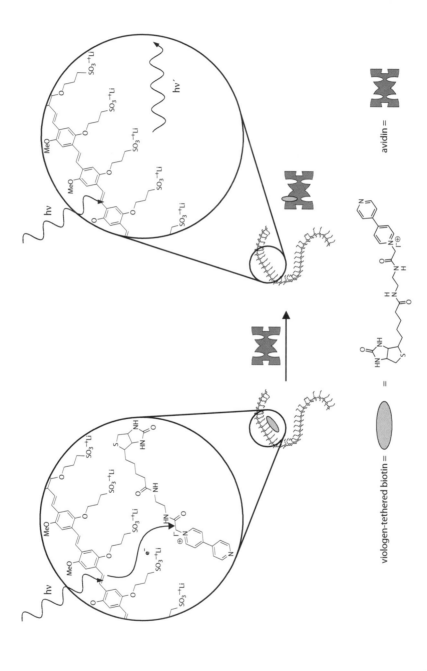

FIGURE 3.3 Cartoon illustrating the quenching of fluorescence of anionic PPV **10** due to the complexation of viologen-tethered biotin (left). When the viologen-tethered biotin is bound to avidin, fluorescence of **10** is restored (right).

3.3.3 COMPLEX FORMATION

Ionic forces provide a mechanism for forming complexes between CPs and other ionic species. This has been used primarily in solution-based sensing applications, but the ability of IFCPs to bind functional ions does, in principle, provide a means of expanding or tuning the functionality of IFCPs in solid-state organic devices. We focus on examples where complex formation goes beyond a predominantly counter-ion effect in some functional way. It is important, however, to keep in mind the ubiquitous presence of non-specific ionic interactions as potential complicating factors.

IFCPs are being widely explored for sensing applications, with recent attention focused on fluorescence-based assays.[19,20,22] CPs are of interest to such assays because a single point of interaction along the polymer chain can quench fluorescence from a large portion of the polymer system, as defined by the exciton diffusion length.[86] The fact that the exciton diffusion length is typically greater than in small molecule fluorophores leads to a relative amplification of the response. As amplification is defined in reference to a small-molecule analogue, the nature of the analogue is important to the definition. Interpretation of apparent amplification factors may require considerations beyond exciton diffusion, such as differences in binding affinity.[20]

In their early work, Chen et al. demonstrated sensing of the glycoprotein avidin using an IFCP complexed with a functional molecular ion.[87] The functional ion they used was a cationically charged viologen tethered to a biotin subunit. In the absence of avidin, the viologen electrostatically binds to the anionic PPV **10**, quenching the fluorescence (Figure 3.3). In the presence of avidin, the tethered biotin binds to avidin and pulls the viologen away from the IFCP. Removal of the viologen quencher leads to a turning on of the fluoresence. The fluorescent response upon removal of the viologen by addition of avidin is amplified because a single viologen can quench the fluorescence of a large section of the IFCP (note there was ~1 viologen per 1000 repeat units of the IFCP). As a functional molecular ion, the biotin derivatized viologen introduces a site for specific interaction with an analyte and a means of transducing that interaction into changes in the fluorescence of the CP. Variations of this technique, where ions are directly involved in the sensing through the association or disassociation of a quencher, have been used extensively for DNA probes[88–90] and protein sensing.[91,92]

10

Direct complexation of IFCPs to single-stranded DNA has also been explored as a means of sensing. Ho et al. report DNA sensing based on colorimetric and fluorescence changes induced by DNA binding to the cationic PTs **11** and **12**.[43,93] The color changes were discussed in terms of planarization of the cationically functionalized PT backbone upon complexation of the anionically charged phosphate backbone of single-stranded DNA. The color was observed to change again when a complementary strand of DNA was added to the solution, which was argued to lead to a twisting of the polymer backbone, and this color change could be related to the number of mismatches in the oligonucleotide hybridization. Changes in the fluorescence of the PT backbone could also be used to detect mismatches in the oligonucleotide hybridization and proved to be much more sensitive than the colorimetric experiments.

11   **12**

Complex formation of IFCPs with bacteria has been explored further in developing antimicrobials. Lu et al. found that the cationic PPE **13** shows biocidal activity against both Gram-negative bacteria and Gram-positive bacteria spores.[94] The biocidal activity was found to be light induced, and further studies observed singlet oxygen production from direct excitation.[95] The singlet oxygen or subsequent reaction products were found to then react with and kill the bacteria. One of the major points of these papers is that the surface charge of the bacteria attracts the positively charged IFCP, thus bringing them in close enough contact for the polymers to be effective.

13

3.3.4 ASSEMBLIES

IFCPs are being utilized in the construction of conjugated polymer assemblies with long-range order. In the simplest examples, IFCPs are adsorbed to templates with ionic surfaces to form extended IFCP structures. One common approach is to employ biological structures as templates. Herland et al. studied biological polyelectrolytes as scaffolds to control molecular architecture.[96] The advantage of using biological polyelectrolytes is that they can be well ordered and of well-defined size. In this case, amyloid fibrils as a scaffold for the assembly of the cationic PTs **14** and **15** yielded luminescent wires on the nanoscale. With this technique, single fibers were employed to make wires with lengths of several microns and a width of 10 nm. The single fibers then assemble into bundles with lengths greater than 100 μm and thicknesses of ~10 μm. In a separate example of biological templates, Liu et al. used a growing fungi as a template for creating CP tubes of the cationic PT **12** on the micron scale.[97] The size of these tubes could be controlled, exploiting the growth patterns of the fungi.

14   **15**

The most widely studied assemblies based on IFCPs are those fabricated from the layer-by-layer (LbL) approach. In LbL assembly, the electrostatic interactions between cationic and anionic functional groups are used to control the growth of a thin polymer film one layer at a time. This technique was first developed for nonconjugated polyelectrolytes by Decher et al. and was shown to provide molecular control over the deposition of the polymer layers (at least in one dimension).[98,99] The technique was subsequently extended to IFCPs and precursors of IFCPs by Ferreira et al.[100] In a typical procedure, a charged substrate is dipped into a polyelectrolyte solution, rinsed, and then dipped into a solution of a polyelectrolyte with opposite charge. This process is repeated iteratively to build up films with as many as 100 layers.[100] Metal-phosphonate chemistry has also been used in the LbL assembly of the phosphonate derivatized PT **16**. Here, the cationic polymer species is replaced with a Zr^{4+} ion, and the low solubility of metal phosphonates is exploited.[55] Using the LbL approach with an IFCP provides not only exquisite control over the structure of the films being deposited, but also the functionality of the CP backbone. The combination of structural control and functionality provided by the LbL technique has led to its use in a number of applications such as polymer light-emitting diodes and electrochromic devices.[101–105]

16

17

Beyond constructing purely conjugated polymer structures, the LbL approach has been used in the construction of nanoscale architectures with other materials. For instance, the technique has been used to control the positioning of donors and acceptors in conjugated polymer photovoltaics. One of the major challenges in organic photovoltaics is that excitons have a very short diffusion length, and so the surface area of the donor-acceptor interface becomes crucial for improving efficiencies. The LbL approach with its precise control over individual layers has been applied in an effort to maximize the donor-acceptor surface area. Li et al. used the LbL approach with the cationic PPV **17** and an anionically functionalized C_{60} compound to form structures with up to nine bilayers.[106] The linear increase in the UV-Vis absorption as the bilayer number increased demonstrated the uniformity of the growth process. X-ray diffraction was used to show that the LbL film was uniform and flat with individual bilayer thicknesses of 21 Å. The films were shown to exhibit photovoltaic behavior when irradiated with white light. Mwaura et al. used the LbL technique to deposit 40 to 60 bilayers composed of the anionic PPEs **18** or **19** (donor layer) and the cationically functionalized C_{60} **20** (acceptor layer), as depicted in Figure 3.4.[107] The N/S ratio measured by XPS showed that there was a 2:1 ratio of the cationically functionalized C_{60} to the anionic PPE **18** instead of the expected 1:1 ratio, which was explained as coming from either aggregation of the C_{60} components or hydrophobic interactions. The power efficiency of the multi-layer at AM1.5 was measured to be between 0.01 and 0.04%, which although low, is an improvement over other photovoltaic cells constructed from the LbL approach.[23]

18

19

20

21

The LbL approach has been combined with templating as a means of further controlling long-range structure. For instance, LbL deposition on sacrificial template particles has been used to fabricate antimicrobial capsules from the PPEs **18** and **21**. Corbitt et al. showed that such IFCP capsules have strong light-induced antimicrobial activity when mixed with solutions of either *Cobetia marina* or *Pseudomonas aeruginosa*.[108] The IFCP capsules were found to attract and trap the bacteria. After the bacteria were concentrated in the IFCP capsules, light exposure generated singlet oxygen, as discussed previously, and killed the trapped bacteria more efficiently than either IFCP coated spheres or spheres with polymer grown on the surface.

LbL has also been combined with silica sphere templating in the design of photonic crystals incorporating IFCPs. The LbL approach provides for nanometer control over the shell thickness to tune the photonic band gap, and the incorporation of an IFCP embeds the functionality of a conjugated polymer into the coating. Kim et al. LbL deposited a shell of the anionic PPE **18** along with poly(dimethyldiallylammonium chloride) onto self-assembled silica spheres to create a luminescent photonic colloidal structure.[109] The

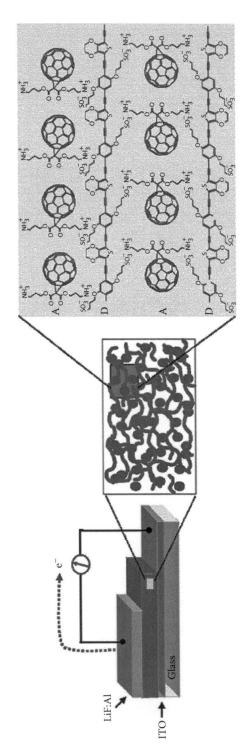

FIGURE 3.4   Schematic of a photovoltaic cell structure fabricated using the LbL approach to deposit alternating layers of the donor **19** (D) and acceptor **20** (A), forming the active material. (From Mwaura, J. K.; Pinto, M. R.; Witker, D.; Ananthakrishnan, N.; Schanze, K. S.; Reynolds, J. R. 2005. Photovoltaic cells based on sequentially adsorbed multilayers of conjugated poly(p-phenylene ethynylene)s and a Water Soluble Fullerene Derivative. *Langmuir 21*: 10119–10126. With permission.)

LbL approach allowed for precise control over the shell thickness and aggregation of the IFCP in the shell, which was used to control absorption and fluorescence properties.

3.4 PHOTOPHYSICS AND QUENCHING
IN UNDOPED SOLID FILMS

Photophysical and quenching processes in CPs have been extensively studied.[110,111] Here, the discussion is limited to aspects related to the presence of ionic functionality. Undoped solid films are the focus because of their relevance to photonic device applications and because elements of the quenching of IFCPs in solution have already been discussed in the previous section.

The simplest way in which ionic functional groups influence photoexcitations is through their effect on molecular structure. As discussed previously, ion-ion interactions can lead to substantial chain aggregation in solution. These studies are relevant to solid films because it has been shown that elements of solution structure can carry through to spin-cast solid films.[112] Indeed, many of the investigations on solution aggregation referenced previously also studied spin-cast films and drew correlations between solution and solid-state structure. In regard to quenching, the majority of work has shown that luminescence quantum yields are reduced in IFCPs relative to their non-ionic counterparts, both in solution and the solid state.[8,9,80,81] They also illustrate that some of the strategies to improve PL quantum yield in solution are also effective at improving the PL quantum yield of solid films. For instance, in the counter-ion dependent studies of Yang et al., the PL quantum efficiency of solid films of the cationic PF **6** cast from water ranged from 5 to 41% depending on counter-ion, with the highest value among the best reported for IFCPs.[80] The low PL quantum yield for solid films of IFCPs contributes to the fact that electroluminescent devices based on these materials show lower efficiencies than closely related blended systems based on CP/polymer-salt complex blends.

Although most studies to date have attributed the role of ions in fluorescence quenching to aspects of molecular structure, it is also possible that ions play a more direct role. Given the strong influence of ionic functionality on chain conformation and interchain interactions, it can be difficult to probe direct interactions between ionic functional groups and optical excitations. Recently, Hodgkiss et al. reported time-resolved PL and transient absorption measurements on films of the cationic PF **22**, which has a relatively low density of ionic functional groups.[113] Their analysis supports the formation of charge transfer states in the IFCP stabilized by interactions with ionic sites. These interactions are not only important because of their effect on diminishing fluorescence quantum yields, but also have implications for applications of CPs in photovoltaics. Exciton lifetime/diffusion and separation into charge carriers are both key to developing efficient CP photovoltaics. The direct influence of ions on the formation of charge transfer states indicates ionic centers may be deleterious in the process of collecting excitons to a charge-separating interface but potentially advantageous at promoting exciton separation at the interface. It is noted that molecular structure, and in particular interchain contacts, is also important to the formation of charge transfer states, as has been studied in solution by Clark et al. for the cationic PPE **23**.[9] As might be expected for nearly all applications of IFCPs, the ability to engineer the precise position of ionic centers is key to controlling function.

22

23

3.5   IONIC TRANSPORT IN UNDOPED IFCPS

As discussed previously, ionic functionality provides a tool for controlling molecular structure. The control of molecular structure through ionic functionality can be used to engineer the properties of IFCPs for electronic and photonic device applications. The role of ions in such applications, however, can go beyond merely control of structure to include a more active role, most simply as ionic charge carriers.

IFCPs are not unique among CPs in containing ions. Conventionally doped CPs contain a substantial density of ionic carriers, but there is also an equal concentration of the electronic carriers that dominate the electrical response because of their higher mobility. Consequently, ion motion in CPs has largely been studied in the context of ion diffusion into or out of polymer thin films during doping.[114,115] What is unique about IFCPs is that they contain ions in their undoped state where there are few, if any, electronic charge carriers. As a result, ion transport will not be swamped out immediately by the transport of electronic carriers.

Ionic conductivity measurements and nearly all device applications of IFCP use ion-blocking electrodes (and redox inactive ions). Such electrodes prevent ions from directly contributing to the long-time, steady-state current. Consequently, ionic conductivity measurements are typically performed using AC techniques with the DC conductivity inferred by extrapolation to as close to zero frequency as the time scale for electrode polarization will allow.[116] As discussed later, the possibility that electronic carriers can be injected into undoped IFCPs can also be a complicating factor and can be minimized by using small amplitude excitation.

24   **25**

There have been few specific studies of ionic conductivity in IFCPs despite its importance to these materials. Lin et al. studied ion transport in amorphous films of anionic PA **24** and cationic PA **25** using small amplitude AC impedance.[63] The stoichiometric density of trifluormethanesulfonate and tetramethylammonium counter-ions, respectively, in the materials is approximately 10^{21} cm^{-3}, and as with nearly all CPs, these materials are well below their glass transition temperatures at room temperature. The ionic conductivity was measured to be in the range of 10^{-11} to 10^{-12} S/cm near room temperature, which is much lower than observed for high performance polymer electrolytes, as typified by polyether-based polymer-salt complexes.[117] The ionic conductivity exhibited Arrhenius temperature dependence with an activiation energy of about 1eV. Such Arrhenius behavior is observed for glassy ionic conductors,[118] and contrasts with the Vogel-Tammann-Fulcher behavior typifying polymer-salt complexes above their glass transition temperatures.[117] The frequency dependence of the ionic conductivity was observed to be similar to that seen in ionic glasses and as typically described by a distributed hopping process.[118] Neher et al. compared the conductivity of sodium and tetradecyltrimethylammonium ions in the anionic PPP **26**.[119] At room temperature, they measured the ionic conductivity at 0.1 Hz to be 7×10^{-12} S/cm and 4×10^{-14} S/cm for Na^+ and tetradecyltrimethylammonium, respectively. Ionic conductivity ultimately places limits on response time of IFCP devices that rely on ion motion, and this response time is often related to the charging of an ionic double layer at a blocking electrode. In the absence of direct measures of the ionic conductivity, the transient behaviors observed in IFCP devices can often be used to compare the relative ionic conductivity of various materials.

$M = -\overset{\oplus}{N}-C_{14}H_{29},\ ^{\oplus}Na,\ or\ ^{\oplus}H$

26   **27**

Efforts to optimize ionic conductivity of IFCPs have not been extensive at this point. IFCPs can be blended or functionalized with polyethers to enhance ionic conductivity and improve the response time of devices, such as polymer light-emitting electrochemical cells, which are discussed more completely later. Oh et al. synthesized the cationic PF **27** that also contained oligoethylene oxide ion-solvating groups attached to the backbone, and they compared these materials to similar polymers **28** and **29** without the ion-solvating groups.[120] They observed that **27** exhibited a time-dependent luminance consistent with the motion of ions (see Figure 3.5), whereas the responses of the polymers **28** and **29** were relatively time independent, consistent with a much lower ionic conductivity. The device with the oligoethylene oxide solvating groups had a lower turn-on voltage, higher luminance, and higher quantum efficiency than a device using similar layers without the solvating groups.

28   **29**

3.6   MIXED IONIC/ELECTRONIC CONDUCTION IN UNDOPED IFCPS

This section focuses on the electrical response of undoped IFCPs because it is in this state where ions most uniquely manifest in transport properties. Ionic conductivity in undoped IFCPs is discussed, followed by the electrical behavior of IFCPs under conditions that also lead to the injection of electronic carriers and hence mixed ionic/electronic transport. Two-electrode, ion-blocking configurations are considered. In such a configuration, it is not possible to change the total ion content of the system, although it can be locally altered. In addition, ion transport cannot contribute to the steady-state current with blocking electrodes. Even so, mobile ions can still contribute to current transients, shape electrostatic potential profiles that influence the injection of electronic charges, and reduce space-charge effects resulting from carrier injection. Mechanisms for the transport of electronic carriers in doped materials is not specifically addressed; the reader is instead referred to the many reviews on transport in doped CPs.[121–125]

3.6.1   CHARGE INJECTION

In principle, a truly intrinsic undoped CP would have limited electronic conductivity due to the absence of electronic charge carriers and the expectation for a relatively low population of thermally generated carriers at room temperature. In reality, many

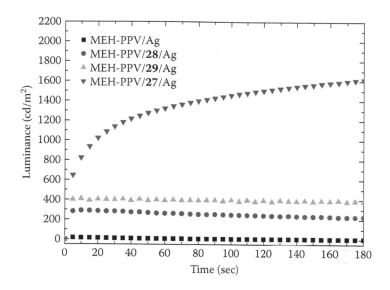

FIGURE 3.5   Luminance as a function of time for MEH-PPV based light-emitting devices without and with injection layers made from IFCPs **27–29**. All were biased at 5 V except the MEH-PPV only device, which was biased at 10 V. (Adapted from Oh, S. H.; Na, S. I.; Nah, Y. C.; Vak, D.; Kim, S. S.; Kim, D. Y. 2007. Novel cationic water-soluble polyfluorene derivatives with ion-transporting side groups for efficient electron injection in PLEDs. *Organic Electronics 8*: 773–783. With permission.)

undoped CPs have some density of intrinsic charged defects that can contribute to the observed electronic conductivity.[126,127] Regardless, it is possible with sufficient electrical bias to inject charge carriers into undoped CPs, leading to a substantial increase in electronic conductivity. The presence of mobile ions in an undoped IFCP strongly affects charge injection by introducing mixed ionic/electronic character, as has been extensively studied in the related field of redox polymers.[114,128]

The importance of mixed ionic/electronic processes in undoped IFCPs to device applications is illustrated by studies on light-emitting devices. Initial work on so-called polymer light-emitting electrochemical cells (PLECs) based on blends of CPs with polymer-salt complexes by Pei et al.[129,130] was shortly followed by studies based on the anionic PPP **26** by Cimrova et al.[131] The key observation was that ion polarization reduced the turn-on voltage for light emission relative to more conventional polymer light-emitting diodes. The low turn-on voltage was argued to be a result of the polarization of ions against the electrodes, leading to large interfacial fields and enhanced charge injection. In both the work on blends and IFCPs, a rise time was reported and attributed to ion motion. In the work of Cimrova et al., a dependence between the turn-on voltage and the counter ion was also reported.[131] A low turn-on voltage of approximately 3 V was observed for Na+, whereas for the large, and much less mobile tetradecyltrimethylammonium ion,[119] the turn-on voltage was observed to be much larger (10 to 15 V) and similar to that observed in light-emitting devices based on non-ionically functionalized PPP analogues.

The dramatic influence ion motion has on electronic transport in IFCPs is readily demonstrated by the time-dependent current in response to a potential step. Figure 3.6 shows current transients resulting from such an experiment on a thin solid film of the cationic PA **25** sandwiched between gold electrodes.[132] There is an initial rise in the current on a time scale consistent with the RC time constant estimated for polarizing an ionic double layer at the metal-electrode interface. This initial rise is followed by a peak and then the current levels off as the ions, injected charges, and associated electrostatic potential profile move toward steady state. The current increases orders of magnitude during the current transient. Associated measurements of the decay from steady state show much higher levels of injected charge carrier density than would be expected within a simple space-charge-limited model, as has been applied to ion-free, highly intrinsic semiconductors or insulators.[133]

The mixed ionic/electronic character of IFCPs will not be apparent if operated at sufficiently high frequencies. Edman et al. studied PLECs based on the cationic PF **4** (X=Br) and used an open-face Au|IFCP|Au geometry with 50 μm electrode gaps.[62] At room temperature, the turn-on voltage for light emission was observed to be 120 V. When the temperature of the device was raised from room temperature to 373 K, the turn-on voltage only decreased to 80 V. When the temperature was raised to 413 K, some 10 K above the polymer's melting point, the turn-on voltage dropped dramatically to 3.5 V. This data is consistent with low ion mobility at temperatures below the melting point of the polymer. The high turn-on voltages at low temperature signaled ion motion was slow on the time scale of the experiment. Above the melting point of the polymer,

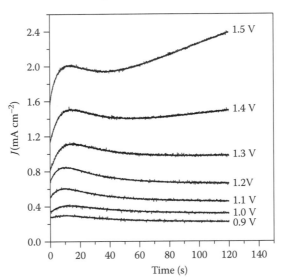

FIGURE 3.6 Time-dependent current density (J) transients produced by a series of separate potential steps, of magnitude indicated, applied to a thin solid film of cationic PA **25** sandwiched between gold electrodes. (Adapted from Cheng, C. H. W.; Lin, F. D.; Lonergan, M. C. 2005. Charge transport in a mixed ionically/electronically conducting, cationic, polyacetylene ionomer between ion-blocking electrodes. *J. Phys. Chem. B 109*: 10168–10178. With permission.)

ion mobility dramatically increases, and ion polarization occurs on the time scale of the experiment leading to a low turn-on voltage. Edman et al. used the strong dependence of the ion mobility on temperature to fabricate so-called frozen junctions.[62] In these junctions, the high-temperature ion distribution that is favorable for charge injection is locked in by lowering the temperature while the sample is held under bias. Such frozen junctions are discussed in greater detail in Chapter 6 of this book.

The mechanistic specifics concerning the mixed ionic/electronic properties described previously have been a topic of considerable interest. Most of the work has been on closely related conjugated polymer/polymer-salt complex blends in the context of PLECs. In principle, the motion of ions in undoped IFCPs can influence charge injection in two primary ways. The first is through ion polarization against blocking electrodes, which leads to large interfacial electric fields that drive charge injection. The second is by charge compensating injected electronic charges, thereby minimizing space-charge effects. The latter method can be seen as the formation of doped sites where injected electronic carriers are charge balanced by the redistribution of ions within the material. The relative importance of these mechanisms and, in a related manner, the details of the electrostatic potential profile within the active layer are discussed further in Chapter 5.

3.6.2   Relation to Conjugated Polymer/Polymer-Salt Complex Blends

IFCPs are often discussed in close parallel to conjugated polymer/polymer-salt complex blends. There are indeed many similarities in the behavior of these two classes of materials, but the fact that IFCPs are single component mixed ionic/electronic conductors makes them distinct from blended systems.[134] IFCPs cannot undergo macroscopic phase separation as has been observed for blended systems,[135] and only one material-solvent interaction has to be optimized for solution processing. Single component IFCP materials are, perhaps, a simpler model system for studying mixed conduction, but control of ion density, type, or the introduction of ion transport promoting functionality requires the synthesis of new materials.

There is a particularly fundamental difference between IFCPs and blends with polymer-salt complexes (but not blends with non-conjugated polyelectrolytes) concerning mixed ionic/electronic transport. Namely, there is only one sign of mobile ion in an IFCP, whereas mobile ions of both signs are present in conjugated polymer/polymer-salt complex blends. It is, in principle, possible to fabricate IFCPs with different signs of bound ionic functionality, but this area has yet to be explored. The presence of only one type of mobile ion has two primary consequences. The first is that the ambipolar diffusion of salt is not possible. Consequently, gradients or abrupt changes in ion concentration are possible with IFCPs (see Section 3.6.3). The second is that ion polarization processes can be more asymmetric than in systems with mobile anions and cations. The application of bias to an IFCP sandwiched between blocking electrodes results in mobile ions being depleted from one side and accumulated at the other. Depending on the details of ion transport, including the maximum obtainable concentration of mobile ions, this asymmetry can lead to differences in the ionic double layer at the two electrodes and hence to an asymmetry in the division of the applied bias between these two interfaces. In some studies of IFCP-based

PLECs, emission could only be observed for one sign of the applied bias and was attributed to an asymmetric polarization process.[119] Such asymmetry has not been noted in all studies of IFCP-based PLECs, however. For instance, Gu et al. report emission from a single layer of the anionic PPV **10**, and a low turn-on voltage was observed for both signs of bias.[136]

3.6.3   UNDOPED POLYMER JUNCTIONS: TRANSPORT LAYERS AND IONIC JUNCTIONS

There are two primary types of interfaces that have been studied concerning the electrical properties of undoped IFCPs. The first are interfaces with ion-blocking metallic or transparent conducting oxide electrodes, which have been discussed throughout because of their ubiquitous presence in IFCP device structures. The second are interfaces between IFCPs and other CPs, and they are the primary topic of this section.

Junctions are at the heart of semiconductor technology.[137] The interface between two materials provides a means of controlling charge transport through offsets in frontier orbital (band edge or HOMO/LUMO) energies and built-in electric fields, which are not possible within the bulk of a homogeneous conductor at steady state. Charge transport and relevant frontier orbital alignments are related to the interfacial structure and bulk material properties such as the Fermi level or the density-of-states profile. This relationship is not unique to ion-containing materials nor does it require ion motion to manifest, but ionic forces can result in particularly strong interactions with any contacting phase. Here, we present examples that illustrate some of the unique properties induced by the presence of ionic functional groups and refer to other sources for more general discussions of interfacial transport at CP interfaces.[138]

In one type of polymer junction, IFCPs are being used to control carrier injection and transport in multilayer light-emitting devices. Efficient light-emitting devices require balancing the injection of electrons and holes, and consequently, material layers are often added in an effort to control this balance.[139] The addition of transport or injection layers can also protect the emissive zone from the electrodes where non-radiative recombination is facile. IFCPs are interesting materials for controlling charge injection because covalently bound ions and their associated counter ions can strongly influence interfacial structure and the spatial profile of the electric field.

30

The earliest motivation for exploring IFCPs in light-emitting devices derived from their solubility characteristics, which facilitated multilayer device fabrication. Hou et al. compared the performance of poly(2-methoxy-5-(2'-ethyl-hexyloxy)-1,4-phenylene vinylene) (MEH-PPV) light-emitting devices with and without the insertion of cationic PF **30** as an electron injection layer between the Al cathode and MEH-PPV emissive layer.[140] They observed that the addition of **30** resulted in orders-of-magnitude improvement in the luminance at substantially lower drive voltages. Other workers reported similar observations on the use of cationic PF derivatives as electron injection layers as well as a dependence on counter anion.[141,142] Recently, Garcia et al. studied multilayer polymer LEDs with MEH-PPV as the emissive layer and the halide salts of **4** as the injection layer.[143] Figure 3.7 shows the change in luminance with bias as a function of counter ion. The luminance vs. bias and luminous efficiency vs. current density were both found to be highest with F⁻ as the counter ion and lowest with I⁻. Wu et al. also compared the cationic PF **31** to its neutral amine functionalized analogue and found that the ionically functionalized polymer led to more efficient devices.[141] These studies did not report time-dependent measurements that might reveal signatures of ion motion.

31

Oh et al. compared the time-dependent luminance of cationic PF-based electron-injecting layers with and without oligoethylene oxide functionality (**27–29**).[120] Inserting a non-oligoethylene oxide containing PF between the emissive layer and Ag cathode resulted in a several-fold increase in luminance efficiency and a largely time-independent luminance. The oligoethylene oxide containing PF **27**, however, exhibited a strong increase in luminance with time, and its long-time efficiency was an order-of-magnitude greater than with the non-oligoethylene oxide containing PFs **28** and **29**. The greater efficiency and time-dependence were argued to be a result of ion motion facilitated by the oligoethylene oxide side chains. This study demonstrates that IFCPs can operate in a manner similar to more traditional injection layers, but that the motion of ions can have a substantial added effect. Hoven et al. also observed a time-dependent luminance with IFCP injection layers of **4** (X=BIm₄⁻) consistent with the ability of ion motion to mediate device performance and specifically the electron injection barrier.[68]

The ability of IFCPs to act as effective injection layers depends on aspects of molecular, interfacial, and electronic structure as well as the redistribution of ions

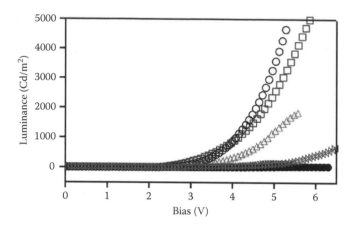

FIGURE 3.7 Luminance vs. bias of multilayer polymer LEDs: ITO/PEDOT:PSS/MEH-PPV/Al (filled circles), ITO/PEDOT:PSS/MEH-PPV/Ba (open circles), and ITO/PEDOT:PSS/MEH-PPV/4X⁻/Al where X = F (squares), Cl (triangles), Br (bowtie), and I (asterisks). PEDOT:PSS = poly(3,4-ethylenedioxythiophene) poly(styrene sulfonate). (From Garcia, A.; Brzezinski, J. Z.; Nguyen, T. Q. 2009. Cationic conjugated polyelectrolyte electron injection layers: effect of halide counterions. *J. Phys. Chem. C 113*: 2950–2954. With permission.)

under bias. It is important, though challenging, to probe the detailed molecular and interfacial structure of IFCPs. Toward this end, Park et al. have performed NEXAFS measurements on thin films of the cationic PF **6** as a function of counter ion and with either MEH-PPV-coated ITO or bare ITO as substrates.[144] They found a substrate-dependent accumulation of the ionic component of the polyelectrolyte at the air-polymer surface. When the IFCP was deposited on the emissive layer (MEH-PPV), they found a higher concentration of ions on the surface than when the IFCP was deposited on ITO. This organization was found to take place on the time scale of the spin casting. If this arrangement were to be maintained through the deposition of the cathode, Park et al. postulated that the orientation of the ions with respect to the surface might be relevant to the injection properties of the IFCP layer.[144]

The electrical characteristics of multilayer CP structures are intimately related to any built-in potential generated by charge transfer equilibration and the division of any applied bias across the various layers/interfaces. In the IFCP/emissive CP structures discussed previously, the electrostatic potential drop across the electrode/IFCP interface relative to that across the emissive layer is likely key. Nevertheless, the IFCP/emissive CP interface may also play an important role. Such polymer-polymer interfaces are not necessarily ion blocking, and the possibility for ion exchange and transport can be likened to similar processes involving electronic carriers at more traditional semiconductor interfaces.

The fact that common IFCPs only have one sign of mobile ion makes possible junctions where there is an asymmetry in ion content. For instance, Cheng et al. studied the current-voltage behavior of an interface between the anionic and cationic PAs **24** and **25** sandwiched between gold electrodes.[145] On the one hand, the polymer interface of this structure bears analogy to the more classic p-n junction with cations and anions replacing the role of holes in the p-type material and electrons in the n-type material, but the analogy breaks down with the blocking metal-polymer interfaces. On the other hand, there is also analogy to purely ionic bipolar membrane structures,[146–149] but in this case, there is no possibility for electronic transport. Cheng et al. observed the strong increase in current upon application of a potential step that is typical for IFCPs and due to ion polarization.[145] However, the magnitude of the increase was observed to be strongly asymmetric with respect to the sign of the applied bias. The ratio of the steady-state current at +3 V to that at –3 V was reported to be approximately 200, and the sign of the bias corresponding to greater current was observed to be consistent with analogy to the p-n junction. In a related junction between ion-functionalized small molecule semiconductors, but with a larger frontier-orbital offset than the PA junction, a photovoltaic effect attributed to an ionic junction has also been observed by Bernards et al.[29]

3.7   DOPING CHEMISTRY

Interest in CPs precipitously increased with the observation that chemical doping of PA could lead to a dramatic increase in conductivity.[1] Recent interest in CPs has shifted more to undoped forms, but the concept of doping remains important to CPs as a means to modulate conductivity, control chemical potential, create asymmetries in carrier type, and because of its connection to charge injection into undoped polymers. CPs can be doped to increase their conductivity either through reduction of the polymer backbone to introduce excess electrons into the π system (n-doping) or oxidation to create holes in the π system (p-doping). The introduction of charge carriers on the polymer backbone requires an accompanying change in ion content for charge neutrality.[150]

The doping of CPs can be achieved chemically or electrochemically. The vast majority of studies highlighting the important distinctions in the doping chemistry of IFCPs have been conducted using electrochemical means, and hence, electrochemical doping is the primary focus herein. The studies described involve thin solid films coated onto an inert electrode in contact with an electrolyte to provide a reservoir of ions. The doping process is also coupled to a complementary redox process at a counter electrode. A number of the concepts discussed would not apply to the solution phase doping of IFCPs.

3.7.1   DOPING MECHANICS AND INTERNALLY COMPENSATED STATES

The mechanics of charge compensation by ions can be very different with IFCPs than with non-ionically functionalized CPs. The key difference is that the doping of a non-ionic CP requires the incorporation of dopant counter-ions. The doping of an IFCP, however, can proceed with the loss of ions built into the polymer structure to form an internally compensated state where charge injected into the polymer backbone is balanced by covalently tethered counter-ions. Figure 3.8 contrasts the p-doping of a conventional CP with that of an anionic IFCP to form an internally compensated state. Note the difference between whether the undoped or doped form contains mobile ions. The situation described by Figure 3.8b, where the density of injected charge is precisely balanced by covalently bound ions, is particularly interesting because it is a pure electronic conductor. This is in contrast to more conventional doped CPs that are mixed ionic/electronic conductors, albeit with a greater electronic than ionic conductivity. The possibility of ionic functional groups becoming built-in dopant counter ions is what led IFCPs to be referred to as self-doped polymers in the original work of Patil et al.[42] The term "internally compensated" is used herein to refer to doped states analogous to Figure 3.8b; the term self-doped is not used because this term often describes IFCPs in general, whether doped or not.

The loss of ions to form an internally compensated state is not the only possible charge compensation mechanism for the doping of an IFCP. It is also possible that charge neutrality be achieved through the incorporation of additional ions. A number of studies have investigated the charge compensation mechanism for the doping of IFCPs using cyclic voltammetry, electrochemical quartz crystal microbalance (EQCM), and compositional analysis.[59,151–153] The general observation is that, where possible, doping occurs with a net loss of counter-ions from the IFCP. Here, "where possible" refers to either the n-doping of a cationic IFCP or the p-doping of an anionic IFCP. For instance, Reynolds et al. used a combination of elemental analysis, chronocoulometry data as a function of cation size, and EQCM to conclude that cation transport dominates the doping chemistry of the anionic PP **32**.[151] The formation of internally compensated states is not surprising given that the incorporation of additional ions within the polymer would amount to the preferential partitioning of a salt in the low dielectric polymer medium relative to the high dielectric solvents typically used for electrochemical doping. There has been, however, a report illustrating a mixed ion flux. Zotti et al. performed a quantitative analysis of EQCM data on the p-doping of the anionic PT **33**, and they found that cation ejection is accompanied by anion incorporation in a 1:1 ratio.[59] Of course, certain doping processes in IFCPs

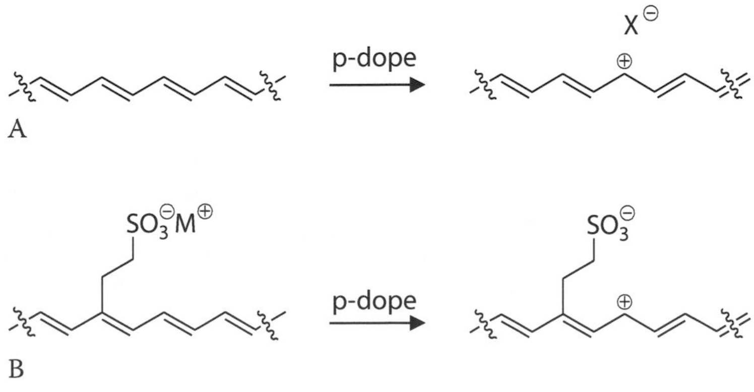

FIGURE 3.8 Comparison of the *p*-doping mechanisms of an anionic IFCP and a conventional CP. (a) *p*-doping of conventional CP. (b) *p*-doping of an anionic IFCP to form an internally compensated state.

must occur with the incorporation of ions: namely, the p-doping of a cationic IFCP, the n-doping of an anionic IFCP, or doping to levels higher than can be compensated by the loss of built-in counterions.

M = K, Li, or H

32   33

Shifting from incorporating an ion of one sign (conventional) to expelling an ion of an opposite sign (as possible with an IFCP) can substantially alter doping kinetics. For instance, Ikenoue et al. reported a tenfold acceleration in the electrochromic switching times accompanying the p-doping of the anionic PT **1**(n = 3, m = H) vs. a non-ionically functionalized PT derivative in HBF_4/CH_3CN electrolyte.[154] The faster kinetics observed with **1** were argued to be due to the more facile transport of the smaller H^+ vs. larger BF_4^- accompanying the p-doping of the anionic vs. nonionic PT, respectively.

3.7.2 Donnan Potentials

The electrochemical doping of CPs as followed by cyclic voltammetry is often used to estimate the frontier-orbital positions of undoped solid-state materials referenced to the vacuum level. This process requires estimating the position of the reference electrode on the vacuum scale[155] and knowledge that the voltammetry data are not complicated by kinetic effects.[156] IFCPs also present an additional challenge due to potential contributions from Donnan potentials.[153,157,158]

IFCPs can act as ion-exchange membranes. When in contact with another electrolyte of differing ion concentration, a transfer of ions will occur due to diffusional gradients resulting in the formation of a junction potential at the IFCP/electrolyte interface. The result is a voltammogram that shifts with electrolyte concentration. For example, shifts of 50 mV per decade over a two to three decade change in electrolyte concentration were observed for the p- and n-doping of anionic and cationic PAs **24** and **25**, respectively.[153] This measurement is close to the ideal expectation of a 60-mV shift per decade change in the salt concentration of the electrolyte. In terms of determining frontier-orbital energies, the Donnan potential can be viewed as an experimental complication, but more fundamentally, it is a manifestation of the fact that interfacial frontier-orbital offsets are a property of the interface that can be difficult to estimate from individual material quantities. The same ion-exchange processes that shift voltammetry data can also occur at solid-state interfaces, thereby influencing frontier-orbital offsets and providing motivation for studying junctions between IFCPs.

3.7.3 Polyelectrolyte Mediated Electrochemistry— Self-Limited and Selective Doping

The counter-ions built into the structure of an IFCP provide a means of using the structure of the conjugated polymer to control doping chemistry. For instance, Gao et al. demonstrated the self-limited electrochemical doping of IFCPs using polyelectrolyte mediated electrochemistry (PMEC).[66] In PMEC, the ions available to support the redox chemistry of a solid film are controlled by restricting their transport into or out of the film. The transport of ions is controlled by using polymeric ions in the supporting electrolyte or solid electroactive film or both.[153,159] In the work of Gao et al., the oxidative p-doping of the anionic PA **24** and the reductive n-doping of the cationic PA **25** were controlled using a supporting electrolyte based on an anionic or cationic polyelectrolyte, respectively.[66] Taking the example of p-doping, the oxidation of an anionic IFCP requires the incorporation of an anion or the loss of a cation, as illustrated in Figure 3.9a. The former was prevented by using tetrabutylammonium polystyrenesulfonate as a supporting electrolyte. The result was a level of p-doping limited by the density of ionic functional groups built into the polymer system (see Figure 3.9b). When compared with that collected in a more conventional electrolyte (dashed line) the cyclic voltammogram in Bu_4NPSS/CH_3CN electrolyte (solid line) shows a much smaller current and correspondingly a lower extent of doping. The ultimate doping level is not controlled by the electrode potential, as in the traditional electrochemical doping of CPs, but by the ion content of the IFCP. An important consequence of this is the suppression of deleterious overdoping processes, which greatly enhances stability in response to large excursions of electrode potential.[153]

The PMEC of IFCPs has also been used to selectively dope one layer of a two-layer structure to form doped/undoped interfaces. For instance, the anionic PA **24** has been selectively p-doped in a two-layer structure that also contains the cationic PA **25** by using tetrabutylammonium polystyrenesulfonate as a supporting electrolyte.[153] It was argued that only the anionic material could be oxidatively p-doped because this process could be supported by the loss of cations from the material, whereas p-doping of the cationic material would require the incorporation of the large polystyrenesulfonate anion.

3.7.4 Doped Polymer Junctions

IFCPs offer distinct opportunities for controlling dopant ion diffusion. The fact that dopant counter ions have some mobility in CPs either as isolated solid films or under doping conditions (e.g., swollen with solvent) complicates the fabrication of interfaces between dissimilarly doped regions. Such interfaces are common in inorganic electronic devices, with a classic example being the silicon p-n junction. In the case of CPs, any diffusion of dopant counter ions would render junctions such as the p-n junction unstable with respect to a bulk redox reaction between the n- and p-type regions, much like the discharge of a shorted battery. This issue has been thoroughly discussed in the context of redox polymers by Buck et al.[160]

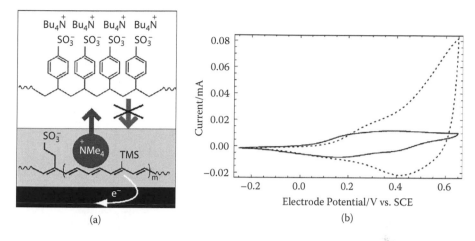

(a)   (b)

FIGURE 3.9   (a) Schematic of the PMEC approach where due to the poly(styrene sulfonate) (PSS) electrolyte, doping can only occur with a loss of a cation from the polymer film resulting in the oxidation level being limited by the available cations in the film. (b) Cyclic voltammogram (CV) of PA **24** solid film collected in 0.1 M Bu$_4$NPSS/CH$_3$CN (solid line). This corresponds to the situation depicted in (a). Also shown is the CV of a PA **24** in 0.1 M Bu$_4$NBF$_4$/CH$_3$CN (dotted line). (Adapted from Gao, L.; Johnston, D.; Lonergan, M. C. 2008. Synthesis and self-limited electrochemical doping of polyacetylene ionomers. *Macromolecules 41*: 4071–4080. With permission.)

A number of approaches have been developed to overcome dopant ion diffusion in the fabrication of junctions between dissimilarly doped CPs. The p-n junction has been a natural focus. A common theme involves the application of bias to a solid film of a mixed ionic/electronic conducting CP system to drive solid-state electrochemical disproportionation, with reduction at one electrode forming an n-type region and oxidation at the other forming a p-type region. The distribution of ions and hence doping is then frozen in by lowering the mobility of the ions either through lowering the temperature[161,162] or by polymerizing the ions[163,164] to increase their molecular weight and lock them in place. The latter approach requires the use of ionic monomers that can undergo photochemically induced polymerization. Neither approach requires the use of IFCPs. Both approaches are discussed further in Chapter 6.

The internally compensated form of IFCPs is particularly interesting in stabilizing interfaces between dissimilarly doped CPs because the dopant ions are bound to the polymer backbone and therefore possess very low mobility. Cheng et al. have exploited this low mobility in the fabrication of a junction between an n-type cationic IFCP and a p-type anionic IFCP.[165] The doped regions were introduced through solid-state electrochemical disproportionation on an undoped bilayer of the IFCPs. The salt liberated from this process was dissolved away by solvent washing to prevent the reverse reaction, leaving the internally compensated p-n junction. The resulting junction exhibited diode behavior in a manner analogous to the silicon p-n junction, as shown in Figure 3.10.

In another approach, Robinson et al. formed an internally compensated p-n junction using a combination of IFCPs and a polyelectrolyte mediated electrochemistry technique.[166] By using both an IFCP and a polyelectrolyte solution in a three-electrode electrochemical technique, control over the ions available to support oxidative and reductive processes was obtained and the dopant density of the individual layers could be independently controlled. This allowed for selective and differential doping of the two polymer layers and the creation of a p-n junction, which exhibited rectification ratios greater than 500 at 1V and a small photovoltaic effect.

3.8   CONCLUDING REMARKS

Ionic functionalization influences many of the key properties of CPs through its impacts on molecular order, the injection and transport of charge, optical excitations, and interactions with other molecules. Consequently, it is an important tool in the development of electronic and photonic devices based on CPs. Continued work on the fundamental mechanisms underlying the physical properties of IFCPs is essential in guiding the rational design of new materials. The development of synthetic and assembly approaches that better control the spatial distribution of ionic centers is expected to open new possibilities. Opportunities exist to exploit both the presence of mobile ions in undoped forms and their absence in internally compensated doped forms. Particularly exciting is the prospect of using ionic functionality to direct charge carriers or control charge recombination and separation in device architectures, much as doping is used to control such processes in more traditional semiconductor devices. More generally, IFCPs are highly complementary to their non-ionically functionalized counterparts, and consequently, it is expected that their study will continue to be an important part of the science and technology of CPs.

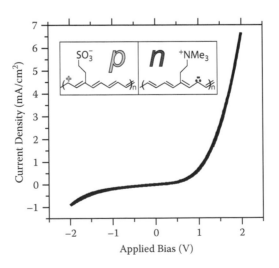

FIGURE 3.10   Current density as a function of applied bias for a *p*-type **24** I *n*-type **25** *p-n* junction. (Adapted from Cheng, C. H. W.; Lonergan, M. C. 2004. A conjugated polymer pn junction. *J. Am. Chem. Soc. 126*: 10536–10537. With permission.)

3.9   ACKNOWLEDGMENTS

M.C.L. acknowledges the National Science Foundation, Division of Materials Research for funding work on the synthesis, electrical, and electrochemical properties of IFCPs, and the Division of Chemical Sciences, Geosciences, and Biosciences, Office of Basic Energy Sciences of the U.S. Department of Energy through grant DE-FG02-07ER15907 for funding work on the photochemical and photophysical properties of IFCPs. The authors acknowledge Lily Robertson and Marcelle Stay for helpful editorial assistance.

REFERENCES

1. Shirakawa, H.; Louis, E. J.; MacDiarmid, A. G.; Chiang, C. K.; Heeger, A. J. 1977. Synthesis of electrically conducting organic polymers—halogen derivatives of polyacetylene. *J. Chem. Soc., Chem. Commun.* 578–580.

2. Bredas, J. L. 1986. Electronic structure of highly conducting polymers. In *Handbook of conducting polymers.* Ed. Skotheim, T. A., New York: Dekker.

3. Grimsdale, A. C.; Chan, K. L.; Martin, R. E.; Jokisz, P. G.; Holmes, A. B. 2009. Synthesis of light emitting conjugated polymers for applications in electroluminescent devices. *Chem. Rev. 109*: 897–1091.

4. Hoven, C. V.; Garcia, A.; Bazan, G. C.; Nguyen, T. Q. 2008. Recent applications of conjugated polyelectrolytes in optoelectronic devices. *Adv. Mater. 20*: 3793–3810.

5. Freund, M. S.; Deore, B.2007. *Self-doped conducting polymers.* Hoboken, NJ.: Wiley.

6. Curtis, M. D.; Cheng, H. T.; Johnson, J. A.; Nanos, J. I.; Kasim, R.; Elsenbaumer, R. L.; Ronda, L. G.; Martin, D. C. 1998. N-methylated poly(nonylbithiazole): A new n-dopable, conjugated poly(ionomer). *Chem. Mater. 10*: 13–16.

7. Langsdorf, B. L.; Zhou, X.; Adler, D. H.; Lonergan, M. C. 1999. Synthesis and characterization of soluble, ionically functionalized polyacetylenes. *Macromolecules 32*: 2796–2798.

8. Nguyen, T. Q.; Schwartz, B. J. 2002. Ionomeric control of interchain interactions, morphology, and the electronic properties of conjugated polymer solutions and films. *J. Chem. Phys. 116*: 8198–8208.

9. Clark, A. P. Z.; Cadby, A. J.; Shen, C. K. F.; Rubin, Y.; Tolbert, S. H. 2006. Synthesis and self-assembly of an amphiphilic poly(phenylene ethynylene) ionomer. *J. Phys. Chem. B 110*: 22088–22096.

10. Radeva, T., Ed. *Physical chemistry of polyelectrolytes.* 2001. Vol. 99. New York: Marcel Dekker.

11. Eisenberg, A.; Rinaudo, M. 1990. Polyelectrolytes and ionomers. *Polym. Bull. 24*: 671–671.

12. Eisenberg, A.; Kim, J. S.1998. *Introduction to ionomers.* New York: Wiley.

13. IUPAC. 1997. *Compendium of chemical terminology*, 2nd ed. (the "Gold Book"). Compiled by A.D. McNaught and A. Wilkinson. Oxford: Blackwell Scientific Publications .

14. Yamamoto, T.; Yoneda, Y.; Maruyama, T. 1992. Ruthenium and nickel-complexes of a pi-conjugated electrically conducting polymer chelate ligand, poly(2,2-bipyridine-5,5'-diyl), and their chemical and catalytic reactivity. *J. Chem. Soc., Chem. Commun.* 1652–1654.

15. Zhu, S. S.; Swager, T. M. 1996. Design of conducting redox polymers: A polythiophene-Ru(bipy)$_3$$^{n+}$ hybrid material. *Adv. Mater. 8*: 497–500.

16. Wolf, M. O.; Wrighton, M. S. 1994. Tunable electron-density at a rhenium carbonyl complex coordinated to the conducting polymer poly 5,5'-(2-thienyl)-2,2'-bithiazole. *Chem. Mater. 6*: 1526–1533.

17. Wang, Q.; Yu, L. P. 2000. Conjugated polymers containing mixed-ligand ruthenium(II) complexes. Synthesis, characterization, and investigation of photoconductive properties. *J. Am. Chem. Soc. 122*: 11806–11811.

18. Pinto, M. R.; Schanze, K. S. 2002. Conjugated polyelectrolytes: Synthesis and applications. *Synthesis-Stuttgart* 1293–1309.

19. Liu, B.; Bazan, G. C. 2004. Homogeneous fluorescence-based DNA detection with water-soluble conjugated polymers. *Chem. Mater. 16*: 4467–4476.

20. Thomas, S. W.; Joly, G. D.; Swager, T. M. 2007. Chemical sensors based on amplifying fluorescent conjugated polymers. *Chem. Rev. 107*: 1339–1386.

21. Huang, F.; Wu, H. B.; Peng, J. B.; Yang, W.; Cao, Y. 2007. Polyfluorene polyelectrolytes and their precursors processable from environment-friendly solvents (alcohol or water) for PLED applications. *Curr. Org. Chem. 11*: 1207–1219.

22. Feng, F.; He, F.; An, L.; Wang, S.; Li, Y.; Zhu, D. 2008. Fluorescent conjugated polyelectrolytes for biomacromolecule detection. *Adv. Mater 20*: 2959–2964.

23. Jiang, H.; Taranekar, P.; Reynolds, J. R.; Schanze, K. S. 2009. Conjugated polyelectrolytes: synthesis, photophysics, and applications. *Angew. Chem. Int. Ed. 48*: 4300–4316.

24. Shimidzu, T.; Ohtani, A.; Iyoda, T.; Honda, K. 1986. Effective adsorption desorption of cations on a polypyrrole polymer anion composite electrode. *J. Chem. Soc., Chem. Commun.* 1415–1417.

25. Miller, L. L.; Zhou, Q. X. 1987. Poly(n-methylpyrrolylium) poly(styrenesulfonate)— a conductive, electrically switchable cation exchanger that cathodically binds and anodically releases dopamine. *Macromolecules 20*: 1594–1597.

26. Hirai, T.; Kuwabata, S.; Yoneyama, H. 1988. Electrochemical behaviors of polypyrrole, poly-3-methylthiophene, and polyaniline deposited on nafion-coated electrodes. *J. Electrochem. Soc. 135*: 1132–1137.

27. Baker, C. K.; Qiu, Y. J.; Reynolds, J. R. 1991. Electrochemically induced charge and mass-transport in polypyrrole/poly(styrenesulfonate) molecular composites. *J. Phys. Chem. 95*: 4446–4452.

28. Ren, X. M.; Pickup, P. G. 1993. Ion-transport in polypyrrole and a polypyrrole polyanion composite. *J. Phys. Chem. 97*: 5356–5362.

29. Bernards, D. A.; Flores-Torres, S.; Abruna, H. D.; Malliaras, G. G. 2006. Observation of electroluminescence and photovoltaic response in ionic junctions. *Science 313*: 1416–1419.

30. Kraft, A.; Grimsdale, A. C.; Holmes, A. B. 1998. Electroluminescent conjugated polymers-seeing polymers in a new light. *Angew. Chem. Int. Ed. 37*: 402–428.

31. Leclerc, M. 2001. Polyfluorenes: twenty years of progress. *J. Polym. Sci. Part A: Polym. Chem. 39*: 2867–2873.

32. Yamamoto, T. 2002. Cross-coupling reactions for preparation of pi-conjugated polymers. *J. Organomet. Chem 653*: 195–199.

33. Babudri, F.; Farinola, G. M.; Naso, F. 2004. Synthesis of conjugated oligomers and polymers: the organometallic way. *J. Mater.Chem. 14*: 11–34.

34. Farinola, G. M.; Babudri, F.; Cardone, A.; Omar, O. H.; Naso, F. 2008. Synthesis of substituted conjugated polymers: Tuning properties by functionalization. *Pure Appl. Chem. 80*: 1735–1746.

35. Shi, S.; Wudl, F. 1990. Synthesis and characterization of a water-soluble poly(p-phenylenevinylene) derivative. *Macromolecules 23*: 2119–2124.

36. Peng, Z.; Xu, B.; Zhang, J.; Pan, Y. 1999. Synthesis and optical properties of water soluble poly(p-phenylenevinylene)s. *Chem. Comm.* 1855–1856.

37. Fujii, A.; Sonoda, T.; Yoshino, K. 2000. Preparation, optical properties and yellow electroluminescence of water soluble poly(p-phenylene vinylene). *Jpn. J. Appl. Phys. 39*: L249–252.

38. Fan, Q. L.; Lu, S.; Lai, Y. H.; Hou, X. Y.; Huang, W. 2003. Synthesis, characterization, and fluorescence quenching of novel cationic phenyl-substituted poly(p-phenylenevinylene)s. *Macromolecules 36*: 6976–6984.
39. Liu, B.; Yu, W. L.; Lai, Y. H.; Huang, W. 2002. Blue-light-emitting cationic water-soluble polyfluorene derivatives with tunable quaternization degree. *Macromolecules 35*: 4975–4982.
40. Stork, M.; Gaylord, B.; Heeger, A. J.; Bazan, G. C. 2002. Energy transfer in mixtures of water soluble oligomers: Effect of charge, aggregation, and surfactant complexation. *Adv. Mater. 14*: 361–366.
41. Huang, F.; Wu, H.; Wang, D.; Yang, W.; Cao, Y. 2004. Novel electroluminescent conjugated polyelectrolytes based on polyfluorene. *Chem. Mater. 16*: 708–716.
42. Patil, A. O.; Ikenoue, Y.; Wudl, F.; Heeger, A. J. 1987. Water-soluble conducting polymers. *J. Am. Chem. Soc. 109*: 1858–1859.
43. Ho, H. A.; Boissinot, M.; Bergeron, M. G.; Corbeil, G.; Dore, K.; Boudreau, D.; Leclerc, M. 2002. Colorimetric and fluorometric detection of nucleic acids using cationic polythiophene derivatives. *Angew. Chem. Int. Ed. 41*: 1548–1551.
44. Ewbank, P. C.; Loewe, R. S.; Zhai, L.; Reddinger, J.; Sauve, G.; McCullough, R. 2004. Regioregular poly(thiophene-3-alkanoic acid)s. Water soluble conducting polymers suitable for chromatic chemosensing in solution and solid state. *Tetrahedron 60*: 11269–11275.
45. Wallow, T. I.; Novak, B. M. 1991. In aqua synthesis of water soluble poly(p-phenylene) derivatives. *J. Am. Chem. Soc. 113*: 7411–7412.
46. Child, A. D.; Reynolds, J. R. 1994. Water-soluble rigid-rod polyelectrolytes: A new self-doped, electroactive sulfonatoalkoxy-substituted poly(p-phenylene). *Macromolecules 27*: 1975–1977.
47. Brodowski, G.; Horvath, A.; Ballauff, M.; Rehahn, M. 1996. Synthesis and intrinsic viscosity in salt-free solution of a stiff-chain cationic poly(p-phenylene) polyelectrolyte. *Macromolecules 29*: 6962–6965.
48. Wittemann, M.; Rehahn, M. 1998. Water-soluble cationic poly-p-phenylene polyelectrolytes with an exceptionally high charge density. *Chem. Comm.*: 623–624.
49. Sundaresan, N. S.; Basak, S.; Pomerantz, M.; Reynolds, J. R. 1987. Electroactive copolymers of pyrrole containing covalently bound dopant ions - poly(pyrrole-co- 3-(pyrrol-1-yl)propanesulphonate). *J. Chem. Soc., Chem. Commun.* 621–622.
50. Pickup, P. G. 1987. Poly-(3-methylpyrrole-4-carboxylic acid)—an electronically conducting ion-exchange polymer. *J. Electroanal. Chem. 225*: 273–280.
51. Havinga, E. E.; Hoeve, W.; Meijer, E. W.; Wynberg, H. 1989. Water-soluble self-doped 3-substituted polypyrroles. *Chem. Mater. 1*: 650–659.
52. Liu, C. H.; Chen, S. H.; Chen, Y. 2006. Synthesis and optical and electrochemical properties of copolymers containing 9,9-dihexylfluorene and 9-dimethylaminopropylcarbazole chromophores. *J. Polym. Sci., Part A: Polym. Chem. 44*: 3882–3895.
53. Zhao, X.; Pinto, M. R.; Hardison, L. M.; Mwaura, J.; Muller, J.; Jiang, H.; Witker, D.; Kleiman, V. D.; Reynolds, J. R.; Schanze, K. S. 2006. Variable band gap poly(arylene ethynylene) conjugated polyelectrolytes. *Macromolecules 39*: 6355–6366.
54. Pinto, M. R.; Kristal, B. M.; Schanze, K. S. 2003. A water-soluble poly(phenylene ethynylene) with pendant phosphonate groups. Synthesis, photophysics, and layer-by-layer self-assembled films. *Langmuir 19*: 6523–6533.
55. Viinikanoja, A.; Lukkari, J.; Aaritalo, T.; Laiho, T.; Kankare, J. 2003. Phosphonic acid derivatized polythiophene: A building block for metal phosphonate and polyelectrolyte multilayers. *Langmuir 19*: 2768–2775.
56. Durben, S.; Dienes, Y.; Baumgartner, T. 2006. Cationic ditheno[3,2-b:2',3'-d]phospholes: a new building block for luminescent conjugated polyelectrolytes. *Org. Lett. 8*: 5893–5896.

57. Forrest, S. R. 2004. The path to ubiquitous and low-cost organic electronic appliances on plastic. *Nature 428*: 911–918.

58. de Boer, B.; Facchetti, A. 2008. Semiconducting polymeric materials. *Polym. Rev. 48*: 423–431.

59. Zotti, G.; Zecchin, S.; Schiavon, G.; Berlin, A.; Pagani, G.; Canavesi, A. 1997. Doping-induced ion-exchange in the highly conjugated self-doped polythiophene from anodic coupling of 4-(4H-cyclopentadithien-4-yl)butanesulfonate. *Chem. Mater. 9*: 2940–2944.

60. Havinga, E. E.; Vanhorssen, L. W.; Tenhoeve, W.; Wynberg, H.; Meijer, E. W. 1987. Self-doped water-soluble conducting polymers. *Polym. Bull. 18*: 277–281.

61. Sonmez, G.; Schwendeman, I.; Schottland, P.; Zong, K. W.; Reynolds, J. R. 2003. N-substituted poly(3,4-propylenedioxypyrrole)s: High gap and low redox potential switching electroactive and electrochromic polymers. *Macromolecules 36*: 639–647.

62. Edman, L.; Liu, B.; Vehse, M.; Swensen, J.; Bazan, G. C.; Heeger, A. J. 2005. Single-component light-emitting electrochemical cell fabricated from cationic polyfluorene: Effect of film morphology on device performance. *J. Appl. Phys. 98*: 044502.

63. Lin, F. D.; Wang, Y. J.; Lonergan, M. 2008. Ion transport in polyacetylene ionomers. *J. Appl. Phys. 104*: 103517.

64. Qin, C.; Cheng, Y.; Wang, L.; Jing, X.; Wang, F. 2008. Phosphonate-functionalized polyflurene as a highly water-soluble iron (III) chemosensor. *Macromolecules 41*: 7798–7804.

65. Langsdorf, B. L.; Zhou, X.; Lonergan, M. C. 2001. Kinetic study of the ring-opening metathesis polymerization of ionically functionalized cyclooctatetraenes. *Macromolecules 34*: 2450–2458.

66. Gao, L.; Johnston, D.; Lonergan, M. C. 2008. Synthesis and self-limited electrochemical doping of polyacetylene ionomers. *Macromolecules 41*: 4071–4080.

67. Shi, W.; Fan, S. Q.; Huang, F.; Yang, W.; Liu, R. S.; Cao, Y. 2006. Synthesis of novel triphenylamine-based conjugated polyelectrolytes and their application as hole-transport layers in polymeric light-emitting diodes. *J. Mater. Chem. 16*: 2387–2394.

68. Hoven, C.; Yang, R.; Garcia, A.; Heeger, A. J.; Nguyen, T. Q.; Bazan, G. C. 2007. Ion motion in conjugated polyelectrolyte electron transporting layers. *J. Am. Chem. Soc. 129*: 10976–10977.

69. He, F.; Feng, F.; Wang, S.; Li, Y. L.; Zhu, D. B. 2007. Fluorescence ratiometric assays of hydrogen peroxide and glucose in serum using conjugated polyelectrolytes. *J. Mater. Chem. 17*: 3702–3707.

70. Fang, Z.; Pu, K. Y.; Liu, B. 2008. Asymmetric fluorescence quenching of dual-emissive porphyrin-containing conjugated polyelectrolytes for naked-eye mercury ion detection. *Macromolecules 41*: 8380–8387.

71. Burrows, H. D.; Lobo, V. M. M.; Pina, J.; Ramos, M. L.; de Melo, J. S.; Valente, A. J. M.; Tapia, M. J.; Pradhan, S.; Scherf, U. 2004. Fluorescence enhancement of the water-soluble poly{1,4-phenylene- 9,9-bis(4-phenoxybutylsulfonate) fluorene-2,7-diyl} copolymer in n-dodecylpentaoxyethylene glycol ether micelles. *Macromolecules 37*: 7425–7427.

72. Gu, Z.; Bao, Y. J.; Zhang, Y.; Wang, M.; Shen, Q. D. 2006. Anionic water-soluble poly(phenylenevinylene) alternating copolymer: high-efficiency photoluminescence and dual electroluminescence. *Macromolecules 39*: 3125–3131.

73. Cheng, F.; Imin, P.; Lazar, S.; Botton, G. A.; Silveira, G.; Marinov, O.; Deen, J.; Adronov, A. 2008. Supramolecular functionalization of single walled carbon nanotubes with conjugated polyelectrolytes and their patterning on surfaces. *Macromolecules 41*: 9869–9874.

74. Masuda, F. 1994. Trends in the development of superabsorbent polymers for diapers. In *Superabsorbent polymers science and technology*. Buchholz, F. L., Pepas, N. A., Eds. Chicago, IL: American Chemical Society, pp. 88–98.

75. Buchholz, F. L.; Graham, A. T. 1998. *Modern superabsorbent polymer technology*. New York: Wiley.
76. Arroyovillan, M. I.; Diazquijada, G. A.; Abdou, M. S. A.; Holdcroft, S. 1995. Poly(N-(3-thienyl)alkanesulfonates)—synthesis, regioregularity, morphology, and photochemistry. *Macromolecules 28*: 975–984.
77. Ortony, J. H.; Yang, R.; Brzezinski, J. Z.; Edman, L.; Nguyen, T. Q.; Bazan, G. C. 2008. Thermophysical properties of conjugated polyelectrolytes. *Adv. Mater. 20*: 298–302.
78. Hara, M., Ed. *Polyelectrolytes science and technology*. 1993. New York: Marcel Dekker.
79. Gaylord, B.; Wang, S.; Heeger, A. J.; Bazan, G. C. 2001. Water-soluble conjugated oligomers: effect of chain length and aggregation on photoluminescence quenching efficiencies. *J. Am. Chem. Soc. 123*: 6417–6418.
80. Yang, R. Q.; Garcia, A.; Korystov, D.; Mikhailovsky, A.; Bazan, G. C.; Nguyen, T. Q. 2006. Control of interchain contacts, solid-state fluorescence quantum yield, and charge transport of cationic conjugated polyelectrolytes by choice of anion. *J. Am. Chem. Soc. 128*: 16532–16539.
81. Garcia, A.; Nguyen, T., Q. 2008. Effect of aggregation on the optical and charge transport properties of an anionic conjugated polyelectrolyte. *J. Phys. Chem. C 112*: 7054–7061.
82. Chen, L. H.; Xu, S.; McBranch, D.; Whitten, D. 2000. Tuning the properties of conjugated polyelectrolytes through surfactant complexation. *J. Am. Chem. Soc. 122*: 9302–9303.
83. Monteserin, M.; Burrows, H. D.; Valente, A. J. M.; Lobo, V. M. M.; Mallavia, R.; Tapia, M. J.; Garcia-Zubiri, I. X.; Di Paolo, R. E.; Macanita, A. L. 2007. Modulating the emission intensity of poly-(9,9-bis(6'-N,N,N-trimethylammonium)hexyl)-fluorene phenylene) bromide through interaction with sodium alkylsulfonate surfactants. *J. Phys. Chem. B 111*: 13560–13569.
84. Knaapila, M.; Almasy, L.; Garamus, V. M.; Pearson, C.; Pradhan, S.; Petty, M. C.; Scherf, U.; Burrows, II. D.; Monkman, A. P. 2006. Solubilization of polyelectrolytic hairy-rod polyfluorene in aqueous solutions of nonionic surfactant. *J. Phys. Chem. B 110*: 10248–10257.
85. Satrijo, A.; Swager, T. M. 2007. Anthryl-doped conjugated polyelectrolytes as aggregation-based sensors for nonquenching multicationic analytes. *J. Am. Chem. Soc. 129*: 16020–16028.
86. Zhou, Q.; Swager, T. M. 1995. Methodology for enhancing the sensitivity of fluorescent chemosensors—energy migration in conjugated polymers. *J. Am. Chem. Soc. 117*: 7017–7018.
87. Chen, L.; McBranch, D. W.; Wang, H. L.; Helgeson, R.; Wudl, F.; Whitten, D. G. 1999. Highly sensitive biological and chemical sensors based on reversible fluorescence quenching in a conjugated polymer. *Proc. Natl. Acad. Sci. 96*: 12287.
88. Gaylord, B.; Heeger, A. J.; Bazan, G. C. 2002. DNA detection using water-soluble conjugated polymers and peptide nucleic acid probes. *Proc. Natl. Acad. Sci. 99*: 10954–10957.
89. Gaylord, B.; Heeger, A. J.; Bazan, G. C. 2003. DNA hybridization detection with water-soluble conjugated polymers and chromophore-labeled single-stranded DNA. *J. Am. Chem. Soc. 125*: 896–900.
90. Xu, H.; Wu, H.; Huang, F.; Song, S.; W, L.; Cao, Y.; Fan, C. 2005. Magnetically assisted DNA assays: high selectivity using conjugated polymers for amplified fluorescent transduction. *Nucleic Acids Res. 33*: e83.
91. Fan, C.; Plaxco, K. W.; Heeger, A. J. 2002. High-efficiency fluorescence quenching of conjugated polymers by proteins. *J. Am. Chem. Soc 124*: 5642–5643.

92. Wang, D.; Gong, X.; Heeger, P. S.; Rininsland, F.; Bazan, G. C.; Heeger, A. J. 2002. Biosensors from conjugated polyelectrolyte complexes. *Proc. Natl. Acad. Sci. 99*: 49–53.

93. Ho, H. A.; Najari, A.; Leclerc, M. 2008. Optical detection of DNA and proteins with cationic polythiophenes. *Acc. Chem. Res. 41*: 166–178.

94. Lu, L. D.; Rininsland, F. H.; Wittenburg, S. K.; Achyuthan, K. E.; McBranch, D. W.; Whitten, D. G. 2005. Biocidal activity of a light-absorbing fluorescent conjugated poly-electrolyte. *Langmuir 21*: 10154–10159.

95. Chemburu, S.; Corbitt, T. S.; Ista, L. K.; Ji, E.; Fulghum, J.; Lopez, G. P.; Ogawa, K.; Schanze, K. S.; Whitten, D. G. 2008. Light-induced biocidal action of conjugated poly-electrolytes supported on colloids. *Langmuir 24*: 11053–11062.

96. Herland, A.; Bjork, P.; Nilsson, K. P. R.; Olsson, J. D. M.; Asberg, P.; Konradsson, P.; Hammerstrom, P.; Inganas, O. 2005. Electroactive luminescent self-assembled bio-organic nanowires: integration of semiconducting oligoelectrolytes within amyloido-genic Proteins. *Adv. Mater. 17*: 1466–1471.

97. Liu, L.; Duan, X.; Liu, H.; Wang, S.; Li, Y. 2008. Microorgansim-based assemblies of luminescent conjugated polyelectrolytes. *Chem. Comm.* 5999–6001.

98. Decher, G.; Hong, J. D. 1991. Buildup of ultrathin multilayer films by a self-assembly process. 1. Consecutive adsorption of anionic and cationic bipolar amphiphiles on charged surfaces. *Makromol. Chem. 46*: 321–327.

99. Decher, G. 1997. Fuzzy nanoassemblies: Toward layered polymeric multicomposites. *Science 277*: 1232–1237.

100. Ferreira, M.; Rubner, M. F. 1995. Molecular-level processing of conjugated polymers. 1. Layer-by-layer manipulation of conjugated polyions. *Macromolecules 28*: 7107–7114.

101. Cutler, C. A.; Bouguettaya, M.; Reynolds, J. R. 2002. PEDOT polyelectrolyte based electrochromic films via electrostatic adsorption. *Adv. Mater. 14*: 684–688.

102. Jain, V.; Sahoo, R.; Mishra, S. P.; Sinha, J.; Montazami, R.; Yochum, H. M.; Heflin, J. R.; Kumar, A. 2009. Synthesis and characterization of regioregular water-soluble 3,4-pro-pylenedioxythiophene derivative and its application in the fabrication of high-contrast solid-state electrochromic devices. *Macromolecules 42*: 135–140.

103. Lukkari, J.; Salomaki, M.; Viinikanoja, A.; Aaritalo, T.; Paukkunen, J.; Kocharova, N.; Kankare, J. 2001. Polyelectrolyte multilayers prepared from water-soluble poly(alkoxythiophene) derivatives. *J. Am. Chem. Soc. 123*: 6083–6091.

104. Onitsuka, O.; Fou, A. C.; Ferreira, M.; Hsieh, B. R.; Rubner, M. F. 1996. Enhancement of light emitting diodes based on self-assembled heterostructures of poly(p-phenylene vinylene). *J. Appl. Phys. 80*: 4067–4071.

105. Baur, J. W.; Kim, S.; Balanda, P. B.; Reynolds, J. R.; Rubner, M. F. 1998. Thin-film light-emitting devices based on sequentially adsorbed multilayers of water-soluble poly(p-phenylene)s. *Adv. Mater. 10*: 1452–1455.

106. Li, H. M.; Li, Y. L.; Zhai, J.; Cui, G. L.; Liu, H. B.; Xiao, S. Q.; Liu, Y.; Lu, F. S.; Jiang, L.; Zhu, D. B. 2003. Photocurrent generation in multilayer self-assembly films fabri-cated from water-soluble poly(phenylene vinylene). *Chem. Eur. J. 9*: 6031–6038.

107. Mwaura, J. K.; Pinto, M. R.; Witker, D.; Ananthakrishnan, N.; Schanze, K. S.; Reynolds, J. R. 2005. Photovoltaic cells based on sequentially adsorbed multilayers of conjugated poly(p-phenylene ethynylene)s and a water soluble fullerene derivative. *Langmuir 21*: 10119–10126.

108. Corbitt, T. S.; Sommer, J. R.; Chemburu, S.; Ogawa, K.; Linnea, I. K.; Lopez, G. P.; Whitten, D.; Schanze, K. S. 2009. Conjugated polyelectrolyte capsules: Light-activated antimicrobial micro "roach motels." *ACS Appl. Mater. 1*: 48–52.

109. Kim, K.; Webster, S.; Levi, N.; Carroll, D. L.; Pinto, M. R.; Schanze, K. S. 2005. Luminescent core-shell photonic crystals from poly(phenylene ethynylene) coated silica spheres. *Langmuir 21*: 5207–5211.

110. Bredas, J. L.; Beljonne, D.; Coropceanu, V.; Cornil, J. 2004. Charge-transfer and energy-transfer processes in pi-conjugated oligomers and polymers: A molecular picture. *Chem. Rev. 104*: 4971–5003.
111. Barbara, P. F.; Gesquiere, A. J.; Park, S. J.; Lee, Y. J. 2005. Single-molecule spectroscopy of conjugated polymers. *Accounts Chem. Res. 38*: 602–610.
112. Nguyen, T. Q.; Kwong, R. C.; Thompson, M. E.; Schwartz, B. J. 2000. Improving the performance of conjugated polymer-based devices by control of interchain interactions and polymer film morphology. *Appl. Phys. Lett. 76*: 2454–2456.
113. Hodgkiss, J. M.; Tu, G.; Seifried, S. A.; Huck, W. T. S.; Friend, R. H. 2009. Ion-induced formation of charge-transfer states in conjugated polyelectrolytes. *J. Am. Chem. Soc. 131*: 8913–8921.
114. Pickup, P. G. 1999. Conjugated metallopolymers. Redox polymers with interacting metal based redox sites. *J. Mater. Chem. 9*: 1641–1653.
115. Doblhofer, K.; Rajeshwar, K. 1998. In *Handbook of conducting polymers,* 2nd ed.; Eds. Skotheim, T. A., Elsenbaumer, R., Reynolds, J. R., New York: Dekker.
116. Barsoukov, E.; MacDonald, J. R.2005. *Impedance spectroscopy: Theory, experiment, and applications.* Hoboken, NJ: Wiley-Interscience.
117. Gray, F. M.1991. Solid polymer electrolytes: Fundamentals and technological applications. NewYork: VCH.
118. Ingram, M. D.; Imrie, C. T.; Ledru, J.; Hutchinson, J. M. 2008. Unified approach to ion transport and structural relaxation in amorphous polymers and glasses. *J. Phys. Chem. B 112*: 859–866.
119. Neher, D.; Gruner, J.; Cimrova, V.; Schmidt, W.; Rulkens, R.; Lauter, U. 1998. Light-emitting devices based on solid electrolytes and polyelectrolytes. *Polym. Adv. Technol. 9*: 461–475.
120. Oh, S. H.; Na, S. I.; Nah, Y. C.; Vak, D.; Kim, S. S.; Kim, D. Y. 2007. Novel cationic water-soluble polyfluorene derivatives with ion-transporting side groups for efficient electron injection in PLEDs. *Organic Electronics 8*: 773–783.
121. Bredas, J. L.; Street, G. B. 1985. Polarons, bipolarons, and solitons in conducting polymers. *Accounts Chem. Res. 18*: 309–315.
122. Moliton, A.; Hiorns, R. C. 2004. Review of electronic and optical properties of semiconducting pi-conjugated polymers: applications in optoelectronics. *Polym. Int. 53*: 1397–1412.
123. Laquai, F.; Wegner, G.; Bassler, H. 2007. What determines the mobility of charge carriers in conjugated polymers? *Phil. Trans. R. Soc. A 365*: 1473–1487.
124. Coropceanu, V.; Cornil, J.; da Silva, D. A.; Olivier, Y.; Silbey, R.; Bredas, J. L. 2007. Charge transport in organic semiconductors. *Chem. Rev. 107*: 926–952.
125. Walzer, K.; Maennig, B.; Pfeiffer, M.; Leo, K. 2007. Highly efficient organic devices based on electrically doped transport layers. *Chem. Rev. 107*: 1233–1271.
126. Gregg, B. A. 2004. Toward a unified treatment of electronic processes in organic semiconductors. *J. Phys. Chem. B 108*: 17285–17289.
127. Gregg, B. A. 2009. Transport in Charged defect-rich π-conjugated polymers. *J. Phys. Chem. C 113*: 5899–5901.
128. Dalton, E. F.; Surridge, N. A.; Jernigan, J. C.; Wilbourn, K. O.; Facci, J. S.; Murray, R. W. 1990. Charge transport in electroactive polymers consisting of fixed molecular redox sites. *Chem. Phys. 141*: 143–157.
129. Pei, Q. B.; Yu, G.; Zhang, C.; Yang, Y.; Heeger, A. J. 1995. Polymer light-emitting electrochemical-cells. *Science 269*: 1086–1088.
130. Pei, Q. B.; Yang, Y.; Yu, G.; Zhang, C.; Heeger, A. J. 1996. Polymer light-emitting electrochemical cells: In situ formation of a light-emitting p-n junction. *J. Am. Chem. Soc. 118*: 3922–3929.

131. Cimrova, V.; Schmidt, W.; Rulkens, R.; Schulze, M.; Meyer, W.; Neher, D. 1996. Efficient blue light emitting devices based on rigid-rod polyelectrolytes. *Adv. Mater. 8*: 585–588.

132. Cheng, C. H. W.; Lin, F. D.; Lonergan, M. C. 2005. Charge transport in a mixed ionically/electronically conducting, cationic, polyacetylene ionomer between ion-blocking electrodes. *J. Phys. Chem. B 109*: 10168–10178.

133. Lampert, M. A.; Mark, P.1970. *Current injection in solids*. New York: Academic Press.

134. Edman, L. 2005. Bringing light to solid-state electrolytes: The polymer light-emitting electrochemical cell. *Electrochim. Acta 50*: 3878–3885.

135. Cao, Y.; Yu, G.; Heeger, A. J.; Yang, C. Y. 1996. Efficient, fast response light-emitting electrochemical cells: Electroluminescent and solid electrolyte polymers with interpenetrating network morphology. *Appl. Phys. Lett. 68*: 3218–3220.

136. Gu, Z.; Shen, Q. D.; Zhang, J.; Yang, C. Z.; Bao, Y. J. 2006. Dual electroluminescence from a single-component light-emitting electrochemical cell, based on water-soluble conjugated polymer. *J. Appl. Polym. Sci. 100*: 2930–2936.

137. Sze, S. M. 2002. *Semiconductor devices physics and technology,* 2nd ed. New York: Wiley.

138. Lonergan, M. 2004. Charge transport at conjugated polymer-inorganic semiconductor and conjugated polymer-metal interfaces. *Annu. Rev. Phys. Chem. 55*: 257–298.

139. Tang, C. W.; Vanslyke, S. A. 1987. Organic electroluminescent diodes. *Appl. Phys. Lett. 51*: 913–915.

140. Hou, L. T.; Huang, F.; Zeng, W. J.; Peng, J. B.; Cao, Y. 2005. High-efficiency inverted top-emitting polymer light-emitting diodes. *Appl. Phys. Lett. 87*: 153509.

141. Wu, H. B.; Huang, F.; Peng, J. B.; Cao, Y. 2005. Efficient electron injection from bilayer cathode with aluminum as cathode. *Synth. Met. 153*: 197–200.

142. Yang, R. Q.; Wu, H. B.; Cao, Y.; Bazan, G. C. 2006. Control of cationic conjugated polymer performance in light emitting diodes by choice of counterion. *J. Am. Chem. Soc. 128*: 14422–14423.

143. Garcia, A.; Brzezinski, J. Z.; Nguyen, T. Q. 2009. Cationic conjugated polyelectrolyte electron injection layers: effect of halide counterions. *J. Phys. Chem. C 113*: 2950–2954.

144. Park, J.; Yang, R. Q.; Hoven, C. V.; Garcia, A.; Fischer, D. A.; Nguyen, T. Q.; Bazan, G. C.; DeLongchamp, D. M. 2008. Structural characterization of conjugated polyelectrolyte electron transport layers by NEXAFS spectroscopy. *Adv. Mater. 20*: 2491–2496.

145. Cheng, C. H. W.; Boettcher, S. W.; Johnston, D. H.; Lonergan, M. C. 2004. Unidirectional current in a polyacetylene hetero-ionic junction. *J. Am. Chem. Soc. 126*: 8666–8667.

146. Lovrecek, B.; Despic, A.; Bockris, J. O. M. 1959. Electrolytic junctions with rectifying properties. *J. Phys. Chem. 63*: 750–751.

147. Lovrecek, B.; Srb, V.; Kunst, B. 1967. Some electrochemical aspects of ion-exchange membrane junctions. *Electrochim. Acta 12*: 905–907.

148. Mauro, A. 1962. Space charge regions in fixed charge membranes and the associated property of capacitance. *Biophys. J. 2*: 179–198.

149. Coster, H. G. L. 1965. A quantitative analysis of voltage-current relationships of fixed charge membranes and associated property of punch-through. *Biophys. J. 5*: 669–674.

150. Heeger, A. J. 2001. Nobel Lecture: Semiconducting and metallic polymers: The fourth generation of polymeric materials. *Rev. Mod. Phys. 73*: 681–700.

151. Reynolds, J. R.; Sundaresan, N. S.; Pomerantz, M.; Basak, S.; Baker, C. K. 1988. Self-doped conducting copolymers—a charge and mass-transport study of poly(pyrrole-co[3-(pyrrol-1-yl)propanesulfonate]). *J. Electroanal. Chem. 250*: 355–371.

152. Ikenoue, Y.; Uotani, N.; Patil, A. O.; Wudl, F.; Heeger, A. J. 1989. Electrochemical studies of self-doped conducting polymers—verification of the cation-popping doping mechanism. *Synth. Met. 30*: 305–319.

153. Lonergan, M. C.; Cheng, C. H.; Langsdorf, B. L.; Zhou, X. 2002. Electrochemical characterization of polyacetylene ionomers and polyelectrolyte-mediated electrochemistry toward interfaces between dissimilarly doped conjugated polymers. *J. Am. Chem. Soc.* *124*: 690–701.
154. Ikenoue, Y.; Tomozawa, H.; Saida, Y.; Kira, M.; Yashima, H. 1991. Evaluation of electrochromic fast-switching behavior of self-doped conducting polymer. *Synth. Met. 40*: 333–340.
155. Reiss, H.; Heller, A. 1985. The absolute potential of the standard hydrogen electrode— a new estimate. *J. Phys. Chem. 89*: 4207–4213.
156. Bard, A. J.; Faulkner, L. R. 1980. *Electrochemical methods: fundamentals and applications.* New York: Wiley.
157. Zhong, C. J.; Doblhofer, K.; Weinberg, G. 1989. The effect of incorporated negative fixed charges on the membrane-properties of polypyrrole films. *Faraday Discuss. 88*: 307–316.
158. Zhong, C. J.; Storck, W.; Doblhofer, K. 1990. The membrane-properties of the electropolymerized conducting pyrrole n-sulfopropyl-pyrrole copolymer. *Ber. Bunsen-Ges. Phys. Chem 94*: 1149–1155.
159. Elliott, C. M.; Redepenning, J. G.; Balk, E. M. 1985. Electronic conductivity of poly tris(5,5'-bis (3-acrylyl-1-propoxy)carbonyl -2,2-bipyridine)ruthenium(0). *J. Am. Chem. Soc. 107*: 8302–8304.
160. Buck, R. P.; Surridge, N. A.; Murray, R. W. 1992. Liquid solid polyelectrolyte diodes and semiconductor analogs. *J. Electrochem. Soc. 139*: 136–144.
161. Gao, J.; Yu, G.; Heeger, A. J. 1997. Polymer light-emitting electrochemical cells with frozen p-i-n junction. *Appl. Phys. Lett. 71*: 1293–1295.
162. Gao, J.; Li, Y. F.; Yu, G.; Heeger, A. J. 1999. Polymer light-emitting electrochemical cells with frozen junctions. *J. Appl. Phys. 86*: 4594–4599.
163. Leger, J. M.; Rodovsky, D. B.; Bartholomew, G. R. 2006. Self-assembled, chemically fixed homojunctions in semiconducting polymers. *Adv. Mater. 18*: 3130–3134.
164. Leger, J. M.; Patel, D. G.; Rodovsky, D. B.; Bartholomew, G. P. 2008. Polymer photovoltaic devices employing a chemically fixed p-i-n junction. *Adv. Funct. Mater. 18*: 1212–1219.
165. Cheng, C. H. W.; Lonergan, M. C. 2004. A conjugated polymer pn junction. *J. Am. Chem. Soc. 126*: 10536–10537.
166. Robinson, S. G.; Johnston, D. H.; Weber, C. D.; Lonergan,M.C. 2010. Polyelectrolyte-mediated electrochemical fabrication of a polyacetylene p-n junction. *Chem. Mater. 22*: 241–246.

4 Electrochemical Biosensors Based on Conducting Polymers

Minh-Chau Pham and Benoît Piro

CONTENTS

4.1 INTRODUCTION

This chapter presents an overview of existing electrochemical biosensors based on conducting polymers, with applications to enzymatic and DNA biosensors. As the concept of conducting polymers has been defined and described in a previous chapter, we will focus directly on the presentation of electrochemical biosensors.

An electrochemical biosensor uses a biomolecule as the recognition element in association with a transducer, which converts the molecular recognition into an electrochemical signal. Electrochemical transducers include amperometric, potentiometric, conductometric, impedimetric, and semiconductor field-effect principles.

Electronically conducting polymers (ECPs) offer many opportunities for coupling the transducer and the bioprobe. Conventional polymers are used as passive matrices or supports to immobilize biomolecules, due to their easy and versatile chemistry. In addition to this, ECPs can be used as active matrices, mediating electron transfer or even transducing a biomolecular recognition into an interpretable electrochemical signal. Due to the numerous complex processes that take place in the electrochemistry of ECPs (electron transfer, electron diffusion, mass transport), a minor perturbation of even one of them may generate an amplified response. This makes ECPs particularly appropriate materials for sensor applications.

4.2   APPLICATION DOMAINS

Conducting polymer-modified electrodes have been investigated for various applications, as enzyme-, DNA-, or immuno-sensors. Excellent reviews have been published on these topics.[1–3] We will focus on the most extensively developed ones, that is, enzyme and DNA biosensors.

4.2.1   ENZYME BIOSENSORS

4.2.1.1   Introduction

Enzymes are proteins that catalyse the transformation of a substrate into a product. When a redox reaction is concerned, the corresponding enzymes are called oxidoreductases. They operate using a redox center, which is commonly buried within the protein envelope (the enzyme shell). This redox center must be recycled by a co-substrate, which undergoes electron transfer (ET). When the enzyme is immobilized on an electrode, ET can be detected or controlled by the electrode. Therefore, the enzyme operates the molecular recognition of the substrate, which is translated into an electric signal. This is the basic principle of an enzyme biosensor. Our chapter focuses only on oxidase enzymes, for which the co-substrate is dioxygen.

4.2.1.2   Principles

Amperometric biosensors based on oxidase enzymes may be classified into three generations depending on the detection principle. In the first generation, the product or the substrate of the enzymatic reaction is directly oxidized or reduced on the electrode. For oxidases, O_2 or H_2O_2 concentrations may be monitored. In the second generation, detection is performed by oxidation or reduction of an artificial redox mediator acting as the electron shuttle. This approach allows reduction of the working potential of the sensor, avoiding interfering reactions. More recently, the third generation suggests the most interesting configuration, where the enzyme is directly "wired" to the electrode so that electron transfer can occur directly between the enzyme and the electrode, without a mediator. Nevertheless, this kind of recycling is restricted to a very few enzymes, for which the redox-active site is not too buried in the protein envelope, and remains "electrically accessible." A review has been published by Schuhmann.[4]

Many oxido-reductase enzyme-based biosensors have been reported. In this chapter, GOD (glucose oxidase), LOD (lactate oxidase), and PyOD (pyruvate oxidase) are investigated more particularly as examples.

GOD is certainly the enzyme most discussed in the literature because of its good stability and its potential applications, for example, in glucose biosensors, which are almost the only biosensor currently proposed on the market. However, LOD and PyOD also have many potential applications. Lactic acid is a chemical compound involved in several biochemical processes. For example, lactate is produced by our organism during a prolonged effort under anaerobic conditions. It can be responsible for acidification of the blood, called acidosis. Lactic acid is also the acid produced by bacteria of the mouth, responsible for the tooth decay known as caries. Finally, lactic acid is used in a variety of foodstuffs as an acidity regulator, in dairy products, for example. For all these reasons, there is a need for lactate biosensors. Pyruvate is also of interest because it can be used illegally as a food supplement in weight-loss programs or for boosting performance in sport. Its monitoring could become essential in the near future.

These three enzymes share the same active site, FAD (flavine adenine dinucleotide). FAD oxidizes the substrate and becomes $FADH_2$, which is afterwards recycled by dioxygen, the natural co-substrate:

$$FADH_2 + O_2 \longrightarrow FAD + H_2O_2$$

In contrast to some other co-enzymes, FAD is immobilized inside the enzyme and cannot diffuse in and out. Its recycling is therefore critical if the co-substrate cannot diffuse.

4.2.1.3 Enzyme Biosensors and Conducting Polymers

Enzymes may be immobilized in or on conducting matrixes, such as conducting polymers (ECPs). Schuhmann[5] was the first to functionalize ECP for enzyme immobilization. In principle, this method ensures enzyme immobilization and ET at the same time. ET may be achieved by the conducting polymer itself, or by incorporation of a redox relay (the mediator) in the polymer matrix, which enables electron hopping and shortens the donor-acceptor distance. Three configurations can be envisaged using ECPs: (1) ECP/immobilized enzyme/mediator in solution, (2) ECP/enzyme without mediator (direct ET), and (3) ECP/enzyme and mediator both immobilized. Configurations (2) and (3) lead to *reagentless biosensors,* where there is no need to add a reactant to the medium.

1. ECP/immobilized enzyme/mediator in solution—There are two conventional procedures for immobilization of enzymes on ECP-modified electrodes. The first consists in trapping the bulky enzyme within the polymer matrix during its electropolymerization. Foulds and Lowe[6] and Umana and Waller[7] demonstrated that GOD can be entrapped in polypyrrole (PPy) during electrosynthesis. Poly(3,4-ethylenedioxythiophene) (PEDOT) was demonstrated to be a very efficient trapping support to construct in one step an amperometric glucose sensor that can operate under flow injection conditions to determine glucose in synthetic serum samples.[8,9] Physical

trapping is the simplest way; however, the catalytic activity of the enzyme is not excellent. The second procedure comprises two steps: first ECP is grown on the electrode surface and then the enzyme is allowed to react chemically with the polymer surface by covalent bonding or by bioaffinity interactions.[10] Besides enhancing the catalytic activity, the main advantage of this sequential procedure is that the electrode can be constructed under optimal conditions adapted to each step. For example, polymerization can be performed in an organic solvent, followed by enzyme immobilization in aqueous solution. Functionalized ECPs may be obtained either by direct electropolymerization of functionalized monomers or by post-functionalization of conventional ECPs.[10–15]

2. ECP/enzyme without mediator (direct ET)—By definition, a conducting polymer should be able to establish an electrical contact between the immobilized enzyme and the electrode, in the configuration of a third-generation biosensor. Nevertheless, it must be noticed that true direct ET has never been evidenced for oxidase enzymes. Some tentatives are published in the literature. De Taxi du Poet et al.[16] reported that GOD was adsorbed on porous membranes containing PPy and indicated direct ET between GOD and the polymer. Van Os et al.[17] also mentioned ET between PPy and LOD. However, controversial data exist about this direct transfer.[18,19] On the contrary, for peroxidases (like horseradish peroxidase, HRP) the redox active group is electrically more accessible than that of the oxidases, and ET is easier.[20,21] This is why bi-enzymatic electrodes, based on co-immobilization of GOD and HRP, for example, were developed.[22] The mechanism is based on a chain reaction, that is, the reaction product of the first enzyme (e.g., H_2O_2) is used as a substrate by the second enzyme. However, H_2O_2 may degrade the electrode, and the biosensor remains sensitive to the O_2 concentration.

3. ECP/enzyme and mediator both immobilized (reagentless biosensors)—Covalent grafting of the redox mediator as a pendant group onto the polymer backbone has been commonly reported with GOD[23–25] and PyOD.[26] A mercaptohydroquinone-based polymer has been reported, with good efficiency. More generally, it has been shown that the quinone group is a good mediator for several enzymes like GOD[27] or PyOD.[28] It appears that quinone derivatives are advantageous substitutes for more classical mediators like Prussian blue, ferrocene, and osmium complexes due to the low working potential.[13,29–35] Indeed, some quinone derivatives present at the same time good thermodynamics (i.e., allow a much lower working potential [WP] than classical mediators) and good kinetics, even when compared to the natural one, dioxygen.[32] Examples concerning PyOD are given in Table 4.1.

Dioxygen is the natural co-substrate for oxidases. This means that O_2 can compete with the artificial mediator M (Scheme 4.1) if the recycling constant for M (k_r) is not significantly higher than that of O_2 (k_r').

TABLE 4.1

Rate Constant (k_r) for the Recycling Step of PyOD, and $E^0{}_{1/2}$ for Each Mediator

Mediator M	$E^0{}_{1/2}$ (mV/SCE)	k_r (M⁻¹s⁻¹)
FcOH	+140	150,000
1,4-BQ	−20	842,000
1,2-NQ	−120	667,000
5-OH-1,4-NQ(JUG)	−250	290,000
O$_2$	—	166,000

Note: Results from Dang, L.A., Haccoun, J., Piro, B., and Pham, M.C. 2006. Reagentless recycling of pyruvate oxidase on a conducting polymer-modified electrode. Electrochim. Acta 51:3934–43.

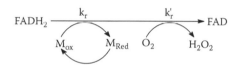

SCHEME 4.1   Schematic representation of the two competitive recycling pathways. The first concerns the artificial mediator M, rate constant k_r; the second concerns the natural mediator, rate constant k_r'.

This is what happens for ferrocene derivatives, for example, and constitutes one of their drawbacks. As shown in Table 4.1, some quinones can compete favorably with dioxygen. For this reason, novel conducting poly(quinones) have been developed recently from amino- and hydroxynaphthoquinone derivatives.[29,36–38] In these cases, quinones are not pendant groups but are partly conjugated with the main polymer backbone, leading to shifts towards more cathodic values for the quinone redox potential. A WP of −0.1 V was achieved, among the lowest ever reported for these enzymes.[26,39,40] This makes these polymers excellent electronic mediators for all FAD-based redox enzymes, such as PyOD and LOD, in de-aerated or even aerated media. Moreover, a low working potential avoids side-oxidation of interfering compounds. In general, the literature does not focus seriously on the working potential problem. However, it is one of the most crucial parameters for an electrochemical sensor.[30–32] The lower it is, the less the electrode will be sensitive to interferents present in biological samples (salicylate, glycine, acetaminophen, ascorbic acid, etc.).

4.2.1.4   Conclusions and Perspectives

Redox enzymes are biomolecules particularly well adapted for integration into electrochemical sensors. Indeed, they involve electroactive species and electron transfer, which can be detected or controlled by an electrode. The crucial point is to achieve the most efficient ET between the enzyme active center and the mediator, at a WP as low as possible. These two objectives may be reached with a redox group having its E° in the cathodic region, but more anodic than that of the redox active group of the enzyme. It has been shown that ECPs cannot play the role of mediators, but they can act as carriers for both enzyme and redox groups and as electron shuttles between these groups. This leads to the construction of *reagentless electrochemical enzyme sensors,* which can operate with no need to add a redox mediator to the solution.

4.2.2   DNA Biosensors

4.2.2.1   Generalities

Detection of a DNA sequence (called the target) is based on its ability to recognize and hybridize with its complementary sequence (called the probe). The procedures for transduction of DNA hybridization are mainly of an optical nature: the target strand is modified with a fluorescent label, hybridized on the probe, and then identified by microfluorescence measurements.[41,42] This technique is reliable and sensitive, but the read-out step is lengthy, as well as the chemical modification of target. There is a need for more simple and fast detection methods; electrochemical techniques are regarded as particularly suitable in this respect.[43] The hybridization should be transduced into a measurable electrical signal. The major difficulty (compared to enzymatic sensors, for example) is that hybridization does not imply any redox reaction. Most of the systems described in the literature imply a chemical modification of the target strands,[44–51] or need addition of a reactant.[52–61] The systems that are capable of direct detection with no need for chemical modification or added reactant,[62–67] namely label-free DNA sensors, are much less frequent.

4.2.2.2   ECP-Based DNA Electrochemical Sensors

ECPs can be used simply as reactive layers to graft DNA probes. However, ECPs can advantageously play the role of transducers because their redox properties can be modulated following hybridization. This makes external reagents unnecessary. Different classes of ECPs were used for DNA sensors: polypyrrole, polyaniline, and polythiophene derivatives are dominant. However, some other ECPs have been investigated.

Polypyrrole and derivatives—Among all ECPs, PPy is the most studied because of its recognized biocompatibility and easy functionalization. In the very first work on this topic, PPy was used only as a substrate for oligonucleotide (ODN) grafting, and not for electrochemical transduction.[68] Extension to electrochemical transduction came later. Hybridization leads to changes in the cyclic voltammograms (CVs) of PPy, that is, a current loss and a potential shift.[69–71]

Redox molecules were also immobilized in ECP as simple redox labels or, recently, as intercalating compounds for an original "fishing" strategy.[72] A review[3] gives an excellent overview of the literature on PPy-based DNA sensors.

Polyaniline and derivatives—Polyaniline (PANi) has been less investigated than PPy, mainly due to its unsatisfactory electrochemical behavior in neutral media. See references in the review.[3]

Polythiophene and derivatives—Works dealing with polythiophene (PT) are more numerous. Some of them are among the first dealing with ECP-based reagentless DNA sensors.[73–75] However, as thiophene is electrooxidized at relatively high potentials (compared to Py or ANi), functionalized thiophene derivatives (bearing -COOH or -NH$_2$ groups) are difficult to electropolymerize and, moreover, these functional groups are likely to react with the radical cations formed during polymerization. Therefore, different approaches were followed in order to lower the oxidation potentials, using oligothiophenes[76] or multi-substituted thiophenes.[77] PEDOT was also described, using the ODN trapping approach.[78,79] See Reference 3 for a more extensive review.

Other ECPs—Some "non-conventional" ECPs have been used to construct electrochemical DNA biosensors: poly(indole),[80] poly(allylamine),[81] or poly(tyramine).[82] Poly(quinone) has been investigated since 2003.[83] A novel copolymer was elaborated: poly(5-hydroxy-1,4-naphthoquinone-co-5-hydroxy-2-acetic acid-1,4-naphthoquinone), poly(JUG-co-JUGA) which allowed direct electrochemical detection. In this copolymer, the quinone group acts as the redox reporter.[66] Some interesting results were obtained concerning the dependence of the signal intensity on the location of the complementary sequence onto the probe strand.[84] The hybridization signal increases with the target length using the same probe.[85] The direct (reagentless) response was compared with that recorded with classical fluorescence experiments.[86] The transduction mechanism was investigated.[85] See Reference 2 for a review.

4.2.2.3   Probe Immobilization Strategies

Short DNA (ODN) sequences used as probes, which are supposed to "fish" for a target sequence, must be immobilized on the electrode surface. ECPs are good candidates for this immobilization due to the diverse possibilities they offer, such as trapping of the probe sequences within the polymer matrix, binding them by affinity interactions, or grafting them via covalent bonds. These methods are detailed in the following.

Trapping—Electrochemical trapping is quite natural with oligonucleotide probes, as they are polyanionic molecules and therefore can play the role of counter-ions in ECP electrodeposition on the electrode, as pioneered by Piro,[36] Wang,[70,87] or in electrooxidation of the monomer. The great advantage of this method, which leads to very high and controllable surface densities of immobilized ODN,[78] is its simplicity. Disadvantages are the poor accessibility of probes to target hybridization and probable irreversible nucleotide damage due to the high potentials used in electropolymerization.

Affinity Immobilization—ODN present less risk than other biomolecules (e.g., proteins, peptides, etc.) of denaturation upon immobilization. However, their immobilization via a bioaffinity complex may be interesting. The avidin-biotin complex is the most usual. This approach was described for the first time with PPy by Cosnier[25] for enzyme immobilization, and Dupont-Fillard[88] for ODN. Another kind of immobilization by the intermediacy of a complex was developed by Thompson,[71] where the terminal phosphate group of an unmodified DNA probe is used as a ligand of a Mg^{2+}-containing polymer.

Covalent Immobilization—ODN probes can be covalently grafted to ECPs. For this, they can be grafted onto a pre-formed polymer (therefore, be grafted only onto the ECP surface), or be functionalized by a monomer and co-electropolymerized with the unmodified monomer (therefore, be grafted on the surface and in the bulk of the ECP). The first work reporting covalent ODN grafting on an ECP was by Livache et al.,[68] although the authors did not develop this approach for electrochemical sensing but rather for optical sensing. Since then, the same group has published numerous improvements still for surface functionalization of optical microarrays. For electrochemical sensing, Garnier et al.[89] were the first to describe an electrochemical procedure to transduce DNA hybridization on ODN-functionalized PPy film,[69] followed by numerous works. More or less at the same period, Bauerle et al.[73,74] developed the same concept with PT. Inspired by the photoactivation/deactivation techniques developed elsewhere, electroactivable ODN-ECP binding has been reported.[90]

PPy and PT are the most studied, due to their easy chemistry, but other kinds of ECPs have been investigated.[80] The nature and redox behavior of the ECPs used for ODN grafting have been scarcely discussed in the literature. However, this may have a noticeable influence in terms of ODN immobilization efficiency or availability toward hybridization. Indeed, as ODN strands are polyanions, they eventually can "stick" electrostatically onto positively charged ECPs[82] instead of being available for hybridization. This is the case for classical p-doped polymers like PANi, PT, or PPy. Negatively charged ECPs capable of cation exchange therefore have been developed to avoid this phenomenon.[36,83]

Surface Probe Density—The probe density on the sensor surface is a crucial parameter for optimizing hybridization. It must be kept in mind that, in general, a high surface concentration does not lead to satisfying results, for two reasons: (1) possible interactions between immobilized single strands, and (2) accessibility to target strands is reduced. The difficulty is to find a balance between a low probe density and a high signal output of the electrode. Some authors have shown that the hybridization efficiency is close to 100% for surface probe densities below 1 pmol.cm^{-2}. In fact, this limit depends on the probe length. A model based on the classical vermicular morphology of polymer chains has been developed to deal with ODN steric hindrance on an electrode surface.[84,85] Table 4.2 summarizes some density limits beyond which interactions between single strands may occur, impeding hybridization.

4.2.2.4 Transduction Strategies and Mechanisms

The electrochemical transduction of DNA hybridization remains a challenge. Indeed, if nucleobases can be electrooxidized (catalytically or not), hybridization itself does not produce any intrinsic electron transfer, so that conversion of hybridization into a measurable electrical signal must be obtained by an extrinsic redox system added in solution or immobilized on the electrode.

For ECP-based DNA sensors, the polymer can be used not only as a supporting matrix for immobilization, but also can play an active role. Indeed, an ECP presents a redox behavior that is due to the polymer backbone electroactivity (PPy, PANi, PT) and due to well-identified redox groups that are present in its structure (ferrocene). Their redox response may be modified by changes in the interactions of the single-stranded probe or double-stranded probe-target with the polymer film. These changes

TABLE 4.2

Gyration Radius R_G, Surface Occupied by a Probe, and Maximum Surface Density of Probes as a Function of the Probe Length (in bases)

Number of Bases	10	20	30	40
R_G/nm	1.20	1.69	2.07	2.39
Probe surface	4.99	9.89	14.8	19.8
Max surf. density/ pmol.cm^{-2}	33.3	16.8	11.2	8.38

Source: From Piro, B., Reisberg, S., Noel, V., and Pham, M.C. 2007. Biosens. Bioelectron. 22:3126–3131. With permission.

can be characterized by amperometry, conductimetry, electrochemical impedance, or simply cyclic voltammetry. Two kinds of sensors, differentiated by their behavior upon hybridization, have been reported: sensors for which hybridization generates a negative signal change, and those for which a positive signal change is detected.

4.2.2.4.1 Negative Signal Changes (Signal-Off Transduction)

The literature most often report amperometric, impedimetric, or conductimetric ECP-based DNA sensors giving a negative signal change upon hybridization.[69,71,74,76,89,91] Such signal-off behavior has often been attributed to polymer reorganization due to the formation of bulky, rigid double-stranded DNA upon hybridization.

In spite of its exhaustive literature, signal-off behavior can be considered as a drawback because it makes the sensor more sensitive to false positives. Indeed, any kind of protein in solution or even non-complementary DNA can adsorb onto the sensor surface and impede the sensor response. That is why positive signal changes (signal-on) are more reliable.

4.2.2.4.2 Positive Signal Changes (Signal-On Transduction)

ECP-based DNA sensors giving a signal rise upon hybridization are scarce. Most of them are indirect ones, relying on a redox reagent added to solution as a probe.[92,93] Others rely on an enzymatic reaction.[94] True direct, reagentless, signal-on transduction was reported for the first time by Pham et al.[83] using a quinone-containing ECP, followed by other work.[84,85,95] Very few other direct detections have been reported.[96]

Quinones are known to be highly sensitive to their chemical environment and, due to their low redox potentials, they lead to an electroactive window in the negative potential domain. This is crucial to avoid side-oxidation of interfering species or even oxidation of nucleobases. The hybridization event can be measured simply by square-wave voltammetry. It was demonstrated that the current decreases after probe grafting and increases after hybridization with the target.[83] No significant change is noticed when a non-complementary strand is added. It was also shown, in other work,[66] that the current changes are mainly due to the faradic process, the capacitive current being more or less constant. Strategies were developed to investigate the steric and the charge effect on the transduction mechanism.[97] It was shown that for a given probe the hybridization signal increases with the target length. It was also demonstrated that very little DNA adsorbs on the polymer film and that the surface concentration of hybrids depends on the target length. A model is proposed to relate the current to the steric hindrance. This allows explaination of the "signal-on" behavior of this sensor and leads to the conclusion that the steric effect is all-important in the electrochemical transduction.[85]

4.2.2.4.3 DNA Substitutes as Probes

Peptide nucleic acids (PNA) are ODN mimics with the same hybridization behavior as ODN, but are neutral. PNA has been investigated with success for electrochemical DNA sensors.[98–100] Original investigations of the transduction mechanism on ECPs were performed using PNA.[97,101] Results confirm the predominance of the steric effect over the charge effect.

4.2.2.5 Conclusions and Perspectives

The numerous electrochemical DNA sensors based on ECPs described in the past years undoubtedly demonstrated that ECPs are powerful materials to transduce the hybridization event. In particular, with adequate functionalization, ECP can lead to reagentless, label-free biosensors capable of making a direct hybridization detection. In perspective, it is certainly interesting to exploit the fact that DNA is, as well as ECP, a polymer, so that they can interact strongly with each other. In conjunction with the development of organic electronics, conjugated oligomers seem promising, as they are compatible with nano-structuration approaches.

REFERENCES

1. McQuade, D.T., Pullen, A.E., Swager, T.M. 2000. Conjugated polymer-based cheical sensors. Chem. Rev. 100:2537–74.
2. Malhotra, B.D., Chaubey, A., Singh, S.P. 2006. Prospects of conducting polymers in biosensors. Anal. Chim. Acta 578:59–74.
3. Peng, H., Zhang, L., Soeller, C., Travas-Sejdic, J. 2009. Conducting polymers for electrochemical DNA sensing. Biomaterials 30:2132–48.
4. Schuhmann, W. 2002. Amperometric enzyme biosensors based on optimized electron-transfer pathways and non-manual immobilization procedures. Rev. Mol. Biotechnol. 82:425–41.
5. Schuhmann, W. 1991. Amperometric substrate determination in flow-injection systems with polypyrrole-enzyme electrodes. Sensors and Actuators, B: Chemical B4:41–9.
6. Foulds, N.C., Lowe, C.R. 1986. Enzyme entrapment in electrically conducting polymers. Immobilization of glucose oxidase in polypyrrole and its application in amperometric glucose sensors. J. Chem. Soc. Faraday Trans. 82:1259–64.
7. Umana, M., Waller, J. 1986. Protein-modified electrodes. The glucose oxidase/polypyrrole system. Anal. Chem. 58:2979–86.
8. Piro, B., Dang, L. A., Pham, M. C., Fabiano, S., Tran-Minh, C. 2001. A glucose biosensor based on modified-enzyme incorporated within electropolymerised poly(3,4-ethylenedioxythiophene) (PEDT) films. J. Electroanal. Chem. 512:101–109.

9. Fabiano, S., Tran-Minh, C., Piro, B., Dang, L. A., Pham, M. C., Vittori, O. 2002. Poly 3,4-ethylenedioxythiophene as an entrapment support for amperometric enzyme sensor. Mat. Sci. Eng. C 21:61–67.

10. Cosnier, S. 2003. Biosensors based on electropolymerized films: New trends. Analytical and Bioanalytical Chemistry 377:507–20.

11. Willner, I., Katz, E., Lapidot, N., Bauerle, P. 1992. Bioelectrocatalyzed reduction of nitrate utilizing polythiophene bipyridinium enzyme electrodes. Bioelectrochem. Bioenerg. 29:29–45.

12. Yang, S.T., Witkowski, A., Hutchins, R.S., Scott, D.L., Bachas, L.G. 1998. Biotin-modified surfaces by electrochemical polymerization of biotinyl-tyramide. Electroanal. 10:58–60.

13. Piro, B., Do, V.A., Le, L.A., Hedayatullah, M., Pham, M.C. 2000. Electrosynthesis of a new enzyme-modified electrode for the amperometric detection of glucose. J. Electroanal. Chem. 486:133–40.

14. Cosnier, S., Fologea, D., Szunerits, S. Marks, R.S. 2000. Poly(dicarbazole-N-hydroxysuccinimide) film: a new polymer for the reagentless grafting of enzymes and redox mediators. Electrochem. Commun. 2:827–31.

15. Situmorang, M., Gooding, J.J., Hibbert, D.B., Barnett, D. 2002. The development of a pyruvate biosensor using electrodeposited polytyramine. Electroanalysis 14:17–21.

16. De Taxis du Poet, P., Miyamoto, S., Murakami, T. K. J., Karube, I. 1990. Direct electron transfer with glucose oxidase immobilized in an electropolymerized poly(N-methylpyrrole) film on a gold microelectrode. Anal. Chim. Acta 235:255–63.

17. Van Os, P.J.H.J., Bult, A., Koopal, C.G.J., van Bannekon, W.P. 1996. Glucose detection at bare and sputtered platinum electrodes coated with polypyrrole and glucose oxidase. Anal. Chim. Acta 335:209–216.

18. Kajiya, Y., Sugai, H., Iwakura, C., Yoneyama, H. 1991. Glucose sensitivity of poly-pyrrole films containing immobilized glucose oxidase and hydroquinonesulfonate ions. Anal. Chem. 63:49–54.

19. Ghindilis, A.L., Atanasov, P., Wilkins, E. 1997. Enzyme-catalyzed direct electron trans-fer. Fundamentals and analytical applications. Electroanalysis 9:661–74.

20. Wollenberger, U., Bogdanovskaya, A., Bobrin, S., Scheller, F., Tarasevich, M.R. 1990. Enzyme electrodes using bioelectrocatalytic reduction of hydrogen peroxide. Anal. Lett. 23:1795–808.

21. Tatsuma, T., Gondaira, M., Watanabe, T. 1992. Peroxidase-incorporated polypyrrole membrane electrodes. Anal. Chem. 64:1183–7.

22. Tatsuma, T., Watanabe, T., Watanabe, T. 1993. Electrochemical characterization of poly-pyrrole bienzyme electrodes with glucose oxidase and peroxidase. J. Electroanal. Chem. 356:245–53.

23. Cosnier, S., Innocent, C., Jouanneau, Y. 1994. Amperometric detection of nitrate via a nitrate reductase immobilized and electrically wired at the electrode surface. Anal. Chem. 66:3198–201.

24. Loughram, M.G., Hall, J.M., Turner, A.P.F. 1996. Development of a pyrroloquinoline quinone (PQQ) mediated glucose oxidase enzyme electrode for detection of glucose in fruit juice. Electroanalysis 8:870–5.

25. Cosnier, S. 1999. Biomolecule immobilization on electrode surfaces by entrapment or attachment to electrochemically polymerized films. A review. Biosens. Bioelectron. 14:443–56.

26. Gajovic, N., Habermüller, K., Warsinke, A., Schuhmann, W., Scheller, F.W. 1999. A pyruvate oxidase electrode based on an electrochemically deposited redox polymer. Electroanalysis 11:1377–83.

27. Arai, G., Masuda, M., Yasumori, I. 1994. Direct electrical response between glucose oxidase and poly(mercapto-p-benzoquinone) films. Bull. Chem. Soc. Jpn. 67:2962–66.

28. Arai, G., Noma, T., Habu, H., Yasumori, I. 1999. Pyruvate sensor based on pyruvate oxidase immobilized in a poly(mercapto-p-benzoquinone) film. J. Electroanal. Chem. 464:143–48.
29. Pham, M.C., Piro, B., Bazzaoui, E.A., Hedayatullah, M., Lacroix, J.C., Novak, P., Haas, O. 1998. Anodic oxidation of 5-amino-1,4-naphthoquinone (ANQ) and synthesis of a conducting polymer (PANQ). Synth. Met. 92:197–205.
30. Haccoun, J., Piro, B., Tran, L.D., Dang, L.A., Pham, M.C. 2004. Reagentless ampero-metric detection of L-Lactate on an enzyme-modified conducting copolymer, poly(5-hydroxy-1,4-naphthoquinone-co-5-hydroxy-3-acetic acid-1,4-naphthoquinone). Biosens. Bioelec. 19:1325–29.
31. Haccoun, J., Piro, B., Noël, V., Pham, M.C. 2006. The development of a reagentless lac-tate biosensor based on a novel conducting polymer. Bioelectrochemistry 68:218–26.
32. Dang, L.A., Haccoun, J., Piro, B., Pham, M.C. 2006. Reagentless recycling of pyruvate oxidase on a conducting polymer-modified electrode. Electrochim. Acta 51:3934–43.
33. Arai, G., Shoji, K., Yasumori, I. 2006. Electrochemical characteristics of glucose oxi-dase immobilized in poly(quinone) redox polymers. J. Electroanal. Chem. 591:1–6.
34. Heller, A., Feldman, B. 2008. Electrochemical glucose sensors and their applications in diabetes management. Chem. Rev. 108:2482–505.
35. Hönes, J., Müller, P., Surridge, N. 2008. The technology behind glucose meters: Test strips. Diabetes Technology & Therapeutics 10:S10–26.
36. Piro, B., Bazzaoui, E.A., Pham, M.C., Novak, P., Haas, O. 1999. Multiple internal reflec-tion FTIR spectroscopic (MIRFTIRS) study of the redox process of poly(5-amino-1,4-naphthoquinone) film in aqueous and organic media. Electrochim. Acta 44:1953–64.
37. Hubert, S., Pham, M.C., Dao, Le H., Piro, B., Nguyen, Q.A., Hedayatullah, M. 2002. A new functionalized conductive polymer poly(2-methyl-5-amino-1,4-naphthoquinone) (PMANQ) with two distinct redox systems. Synth. Met. 128:67–81.
38. Pham, M.C., Hubert, S., Piro, B., Maurel, F., Dao, L. H., Takenouti, H. 2004. Investigations of the redox process of conducting poly(2-methyl-5-amino-1,4-naphtho-quinone) (PMANQ) film interactions of quinone-amine in the polymer matrix. Synth. Met. 140:183–97
39. Gajovic, N., Binyamin, G., Warsinke, A., Frieder, W., Heller, A. 2000. Operation of a miniature redox hydrogel-based pyruvate sensor in undiluted deoxygenated calf serum. Anal. Chem. 72:2963–68.
40. Revzin, A.F., Sirkar, K., Simonian, A., Pishko, M.V. 2002. Glucose, lactate, and pyru-vate biosensor arrays based on redox polymer/oxidoreductase nanocomposite thin-films deposited on photolithographically patterned gold microelectrodes. Sensors and Actuators B 81:359–68.
41. Fodor, S., Read, J.L., Pirrung, M.C., Stryer, L., Tsai, L.A., Solas, D. 1991. Light-directed, spatially addressable parallel chemical synthesis. Science 251:767–73.
42. Piunno, P.A.E., Krull, U.J., Hudson, R.H.E., Damha, M.J., Cohen, H. 1994. Fiber optic biosensor for fluorimetric detection of DNA hybridization. Anal. Chim. Acta 288:205–14.
43. Szunerits, S., Bouffier, L., Calemczuk, R. et al. 2005. Comparison of different strategies on DNA chip fabrication and DNA-sensing: Optical and electrochemical approaches. Electroanalysis 17:2001–17.
44. De Lumley-Woodyear, T., Campbell, C.N., Heller, A. 1996. Direct enzyme-ampli-fied electrical recognition of a 30-base model oligonucleotide. J. Am. Chem. Soc. 118:5504–5.
45. Azec, F., Grossiord, C., Joannes, M., Limoges, B., Brossier, P. 2000. Hybridization assay at a disposable electrochemical biosensor for the attomole detection of amplified human cytomegalovirus DNA. Anal. Biochem. 284:107–13.
46. Wang, J., Kawde, A.N., Musameh, M., Rivas, G. 2002. Dual enzyme electrochemical coding for detecting DNA hybridization. Analyst 127:1279–82.

47. Patolsky, F., Weizmann, Y., Willner, I. 2002. Redox-active nucleic-acid replica for the amplified bioelectrocatalytic detection of viral DNA. J. Am. Chem. Soc. 124:770–2.

48. Palecek, E., Fojta, M., Jelen, F. 2002. New approaches in the development of DNA sensors: Hybridization and electrochemical detection of DNA and RNA at two different surfaces. Bioelectrochem. 56:85–90.

49. Fojta, M., Havran, L., Billova, S., Kostecka, P., Masarik, M., Kizek, R. 2003. Two-surface strategy in electrochemical DNA hybridization assays: Detection of osmium-labeled target DNA at carbon electrodes. Electroanalysis 15:431–40.

50. Anne, A., Bouchardon, A., Moiroux, J. 2003. 3'-Ferrocene-labeled oligonucleotide chains end-tethered to gold electrode surfaces: Novel model systems for exploring flexibility of short DNA using cyclic voltammetry. J. Am. Chem. Soc. 125:1112–13.

51. Dominguez, E., Rincon, O., Narvaez, A. 2004. Electrochemical DNA sensors based on enzyme dendritic architectures: An approach for enhanced sensitivity. Anal. Chem. 76:3132–8.

52. Molinier-Jumel, C., Malfoy, B., Reynaud, J.A., Aubel-Sadron, G. 1978. Electrochemical study of DNA-anthracyclines interaction. Biochem. Biophys. Res. Comm. 84:441–9.

53. Carter, M.T., Bard, A.J. 1987. Voltammetric studies of the interaction of tris(1,10-phenanthroline)cobalt(III) with DNA. J. Am. Chem. Soc. 109:7528–30.

54. Millan, K.L., Mikkelsen, S.R. 1993. Sequence-selective biosensor for DNA based on electroactive hybridization indicators. Anal. Chem. 65:2317–23.

55. Hashimoto, K., Ito, K., Ishimori, Y. 1994. Sequence-specific gene detection with a gold electrode modified with DNA probes and an electrochemically active dye. Anal. Chem. 66:3830–3.

56. Palanti, S., Marrazza, G., Mascini, M.1996. Electrochemical DNA probes. Anal. Lett. 29:2309–31.

57. Oliveira-Brett, A.M., Macedo, T.R.A., Raimundo, D., Marques, M.H., Serrano, S.H.P. 1998. Voltammetric behavior of mitoxantrone at a DNA-biosensor. Biosens. Bioelec. 17:861–7.

58. Marrazza, G., Chiti, G., Mascini, M., Anichini, M. 2000. Detection of human apolipoprotein E genotypes by DNA electrochemical biosensor coupled with PCR. Clin. Chem. 46:31–7.

59. Katz, E., Willner, I., Wang, J. 2004. Electroanalytical and bioelectroanalytical systems based on metal and semiconductor nanoparticles. Electroanalysis 16:19–44.

60. Del Pozo, M.V., Alonso, C., Pariente, F., Lorrenzo, E. 2005. Electrochemical DNA sensing uing osmium complexes as hybridization indicators. Biosens. Bioelec. 20:1549–58.

61. Gorodetsky, A.A., Barton, J.K. 2006. Electrochemistry using self-assembled DNA monolayers on highly oriented pyrolytic graphite. Langmuir 22:7917–22.

62. Palecek, E. 1988. Adsorptive transfer stripping voltammetry: Determination of nanogram quantities of DNA immobilized at the electrode surface. Anal. Biochem. 170:421–31.

63. Kelley, S.O., Boon, E.M., Barton, J.K., Jackson, N.M., Hill, M.G. 1999. Single-base mismatch detection based on charge transduction through DNA. Nucl. Acids. Res. 27:4830–37.

64. Palecek, E. 2002. Past, present and future of nucleic acids electrochemistry. Talanta 56:809–19.

65. Cloarec, J.P., Deligianis, N., Martin, J.R. et al. 2002. Immobilization of homooligonucleotide probe layers onto Si/SiO$_2$ substrates: Characterization by electrochemical impedance measurements and radiolabeling. Biosens. Bioelec. 17:405–12.

66. Piro, B., Haccoun, J., Pham, M.C., Tran, L.D., Rubin, A., Perrot, H., Gabrielli, C. 2005. Study of the DNA hybridization transduction behavior of a quinone-containing electroactive polymer by cyclic voltammetry and electrochemical impedance spectroscopy. J. Electroanal. Chem. 577:155–65.

67. Wong, E.L.S., Mearns, F.J., Gooding, J.J. 2005. Further development of an electrochemical DNA hybridization biosensor based on long-range electron transfer. Sensors and Actuators B: Chemical 111–112:515–21.

68. Livache, T., Roget, A., Dejean, E., Barthet, C., Bidan, G., Teoule, R. 1994. Preparation of a DNA matrix via an electrochemically directed copolymerization of pyrrole and oligonucleotides bearing a pyrrole group. Nucl. Acids. Res. 22:2915–21.

69. Korri-Youssoufi, H., Garnier, F., Srivastava, P., Godillot, P., Yassar, A. 1997. Toward bioelectronics: Specific DNA recognition based on an oligonucleotide-functionalized polypyrrole. J. Am. Chem. Soc. 119:7388–89.

70. Wang, J., Jiang, M., Fortes, A., Mukherjee, B. 1999. New label-free DNA recognition based on doping nucleic-acid probes within conducting polymer films. Anal. Chim. Acta 402:7–12.

71. Thompson, L.A., Kowalik, J., Josowicz, M., Janata, J. 2003. Label-free DNA hybridization probe based on a conducting polymer. J. Am. Chem. Soc. 125:324–5.

72. Cosnier, S., Ionescu, R.E., Herrmann, S., Bouffier, L., Demeunynck, M., Marks, R.S. 2006. Electroenzymatic polypyrrole-intercalator sensor for the determination of West Nile virus cDNA. Anal. Chem. 78:7054–7.

73. Bauerle, P., Emge, A. 1998. Specific recognition of nucleobase-functionalized polythiophenes. Adv. Mater. 10:324–30.

74. Emge, A., Bauerle, P. 1999. Molecular recognition properties of nucleobase-functionalized polythiophenes. Synth. Met. 102:1370–3.

75. Shim, Y.B., Lee, T.Y. 2001. Direct DNA hybridization detection based on the oligonucleotide-functionalized conductive polymer. Anal. Chem. 73:5629–32.

76. Peng, H., Zhang, L., Spires, J., Soeller, C., Travas-Sejdic, J. 2007. Synthesis of a functionalized polythiophene as an active substrate for a label-free electrochemical genosensor. Polymer 48:3413–9.

77. Gautier, C., Cougnon, C., Pilard, J.F., Casse, N. 2006. Label-free detection of DNA hybridization based on EIS investigation of conducting properties of functionalized polythiophene matrix. J. Electroanal. Chem. 587:276–83.

78. Piro, B., Pham, M.C., Le Doan, T. 1999b. Electrochemical method for entrapment of oligonucleotides in polymer-coated electrodes. J. Biomed. Mat. Res. 46:566–72.

79. Krishnamoorthy, K., Gokhale, R.S., Contractor, A.Q., Kumar, A. 2004. Novel label-free DNA sensors based on poly(3,4-ethylenedioxythiophene). Chem. Commun. 7:820–1.

80. Nie, G., Zhang, Y., Guo, Q., Zhang, S. 2009. Label-free DNA detection based on a novel nanostructured conducting poly(indole-6-carboxylic acid) films. Sens. Actuators B 139:592–7.

81. Sato, H., Anzai, J.I. 2006. Preparation of layer-by-layer thin films composed of DNA and ferrocene-bearing poly(amine) and their redox properties. Biomacromol. 7:2072–6.

82. Tran, L. D., Piro, B., Pham, M. C., Ledoan, T., Angiari, C., Dao, L. H., Teston, F. 2003. A polytyramine film for covalent immobilization of oligonucleotides and hybridization. Synth. Met. 139:251–62.

83. Pham, M.C., Piro, B., Tran, L.D., Ledoan, T., Dao, L. H. 2003. Direct electrochemical detection of oligonucleotide hybridization on poly(5-hydroxy-1,4-naphthoquinone-co-5-hydroxy-3-thioacetic acid-1,4-naphthoquinone) film. Anal. Chem. 75:6748–52.

84. Reisberg, S., Piro, B., Noel, V., Pham, M.C. 2005. DNA electrochemical sensor based on conducting polymer: Dependence of the "signal-on" detection on the probe sequence localization. Anal. Chem. 77:3351–6.

85. Piro, B., Reisberg, S., Noel, V., Pham, M.C. 2007. Investigations of the steric effect on electrochemical transduction in a quinone-based DNA sensor. Biosens. Bioelectron. 22:3126–31.

86. Reisberg, S., Piro, B., Noel, V., Pham, M.C. 2006. Selectivity and sensitivity of a reagent-less electrochemical DNA sensor studied by square wave voltammetry and fluorescence. Bioelectrochemistry 69:172–9.

87. Wang, J., Jiang, M. 2000. Toward genoelectronics: Nucleic acid doped conducting polymers. Langmuir 16:2269–74.

88. Dupont-Fillard, A., Roget, A., Livache, T., Billon, M. 2001. Reversible oligonucleotide immobilization based on biotinylated polypyrrole film. Anal. Chim. Acta 449:45—50.

89. Garnier, F. Korri-Youssoufi, H., Srivastava, P., Mandrand, B., Delair, T. 1999. Toward intelligent polymers: DNA sensors based on oligonucleotide-functionalized polypyrroles. Synth. Met. 100:89–94.

90. Pilard, J.F., Cougnon, C., Rault-Berthelot, J., Berthelot, A., Hubert, C., Tran, K. 2004. Elaboration of a novel conducting polymer substituted by a N,N-diethylbenzenesulfonamide group capable of S-N cathodic cleavage in both aqueous and nonaqueous media. J. Electroanal. Chem. 568:195–201.

91. Zhu, N., Chang, Z., He, P., Fang, Y. 2006. Electrochemically fabricated polyaniline nanowire-modified electrode for voltammetric detection of DNA hybridization. Electrochim. Acta 51:3758–62.

92. Le Floch, F., Ho, H.A., Harding-Lepage, P., Bedard, M., Neagu-Plesu, R., Leclerc, M. 2005. Ferrocene-functionalized cationic polythiophene for the label-free electrochemical detection of DNA. Adv. Mater 17:1251–4.

93. Fang, B., Jiao, S., Li, M., Qu, Y., Jiang, X. 2008. Label-free electrochemical detection of DNA using ferrocene-containing cationic polythiophene and PNA probes on nanogold modified electrodes. Biosens. Bioelectron. 23:1175–9.

94. Ionescu, R.E., Herrmann, S., Cosnier, S., Marks, R.S. 2006. A polypyrrole cDNA electrode for the amperometric detection of the West Nile Virus. Electrochem. Commun. 8:1741–8.

95. Acevedo, D.F., Reisberg, S., Piro, B., Peralta, D.O., Miras, M.C., Pham, M.C., Barbero, C.A. 2008. Fabrication of an interpenetrated network of carbon nanotubes and electroactive polymers to be used in oligonucleotide biosensing. Electrochimica Acta 53:4001–6.

96. Komarova, E., Aldissi, M., Bogomolova, A. 2005. Direct electrochemical sensor for fast reagent-free DNA detection. Biosens. Bioelectron. 21:182–9.

97. Reisberg, S., Dang, L.A., Piro, B., Noel, V., Nielsen, P.E., Pham, M.C. 2008. Label-free DNA electrochemical sensor based on a PNA-functionalized conductive polymer. Talanta 76:206–10.

98. Wang, J., Pelecck, E., Nielsen, P.E. et al. 1996. Peptide nucleic acid probes for sequence-specific DNA biosensors. J. Am. Chem. Soc. 118:7667–70.

99. Ozkan, D. et al. 2002. Electrochemical detection of hybridization using peptide nucleic acids and methylene blue on self-assembled alkanethiol monolayer modified gold electrodes. Electrochem. Commun. 4:796–802.

100. Aoki, H., Tao, H. 2007. Label- and marker-free gene detection based on hybridization-induced conformational flexibility changes in a ferrocene-PNA conjugate probe. Analyst 132:784–91.

101. Reisberg, S., Piro, B., Noel, V., Nguyen, T.D., Nielsen, P.E., Pham. M.C. 2008. Investigation of the charge effect on the electrochemical transduction in a quinone-based DNA sensor. Electrochim. Acta 54:346–51.

5 The Light-Emitting Electrochemical Cell

Ludvig Edman

CONTENTS

5.1 INTRODUCTION

The light-emitting electrochemical cell (LEC) can conveniently and broadly be defined as a device that generates light with reversible electrochemical means within a material that allows for both electron *and* ion transport. The concept of employing such electrochemical cells for the generation and study of light-emitting processes was conceived and demonstrated in the 1960s[1–4] and the field has thereafter flourished and expanded, as evidenced, for example, by a number of excellent reviews on the topic.[5–12] The main focus of this chapter is on relatively recent high-performance solid-state LECs, which have begun to attract commercial interest in specialty lighting and display applications,[13] but we find it appropriate to begin in Section 5.2 with a historical overview of the research work in related LEC fields that laid the groundwork for today's state-of-the-art devices. Thereafter, in Section 5.3, the main types of solid-state LEC devices are introduced, and in Section 5.4 the complex and intensely debated operational mechanism of such solid-state LECs is discussed. In Sections 5.5 and 5.6, a number of key parameters of solid-state LECs in the context of usefulness in applications—the turn-on time, the power conversion efficiency, and the operational lifetime—are reviewed. The chapter ends with a conclusion in Section 5.7, where the opportunities and challenges for these fascinating devices in the future are highlighted.

5.2 HISTORICAL BACKGROUND

The first LEC devices—here termed *liquid LECs* for simplicity—utilized a solution comprising one (or more) electrochemically active and luminescent organic compounds and an inert electrolyte as the liquid active material in a two (or three) electrode setup to create light in a process commonly termed solution electrogenerated chemiluminescence.[4,14,15] In one conceptually straightforward manifestation, the same organic compound participates in physically separated electrochemical reactions at the two electrode/active material interfaces: it is oxidized at the positive anode and reduced at the negative cathode to form cationic and anionic radicals, respectively. These radicals then move toward each other, primarily driven by convection in their liquid environment,[15,16] and meet in the bulk where they react chemically to form electronically excited states, which subsequently can decay to the ground state under the emission of light. Moreover, in concentrated electrolyte solutions (as commonly employed in LEC devices), it is further plausible that the cationic and anionic radicals to a significant extent are electrostatically "neutralized" via the formation of coordination complexes with counter ions from the inert electrolyte.[5] The reaction sequence under such conditions, with R being the organic compound, A⁻ and C⁺ being the anion and the cation in the inert electrolyte, respectively, and $\cdots$ indicating the formation of a coordination complex, is thus:

$$R - e^- + A^- \rightarrow R^{+\bullet} \cdots A^- \quad \text{(oxidation at anode)} \tag{5.1}$$

$$R + e^- + C^+ \rightarrow R^{-\bullet} \cdots C^+ \quad \text{(reduction at cathode)} \tag{5.2}$$

$$R^{+\bullet} \cdots A^- + R^{-\bullet} \cdots C^+ \rightarrow R^* + R + A^- + C^+ \quad \text{(excited state formation)} \tag{5.3}$$

$$R^* \rightarrow R + hf \quad \text{(light emission)} \tag{5.4}$$

The most commonly employed type of organic compound in liquid LECs is the so-called conjugated "small molecules," where a few representative examples are $Ru(bpy)_3^{2+}(X^-)_2$ (where bpy is 2,2'-bipyridine and X⁻ typically is a molecular anion such as ClO_4^- or PF_6^-) and rubrene,[17–21] but a small number of functional devices with conjugated polymers as the dissolved organic compound have also been reported.[22–25]

Electrochemically active organic materials, both in the form of conjugated small molecules and polymers, can be deposited as a mechanically stabile thin solid film on an electrode surface with a range of different techniques, for example, in the case of small molecules, via immobilization in a solid matrix or *in-situ* polymerization, or in the case of polymers, via spin-coating or printing. If such a modified electrode with a solid organic film on top is part of an appropriate three-electrode setup immersed in an inert electrolyte solution (in which the film is insoluble), then it is a straightforward task to establish the redox potentials of the organic compound in the solid state with the cyclic voltammetry (CV) technique.[26–28] Moreover, the first oxidation and reduction potentials recorded in a CV scan can be used for the *estimation* of the important highest occupied molecular orbital (HOMO) and lowest unoccupied molecular orbital (LUMO) values, respectively, of the organic compound. It is notable that an influx of ions into the solid organic film from the liquid electrolyte is necessary during the redox processes to allow for efficient electronic charge injection via electric double layer formation and for the preservation of electroneutrality in the bulk, and that this critical passage of an ionic current through the (in some cases μm-thick) organic film is facilitated by the characteristic porous and soft nature of solution-processed organic materials.

If the organic compound on top of the modified electrode is reasonably luminescent (i.e., fluorescent or phosphorescent), in addition to allowing for reversible reduction *and* oxidation, then the same setup as utilized for CV measurements can form the basis for the herein termed *hybrid LEC* devices, for which the redox reactions takes place in the solid-state film while the charge compensating counter-ions initially are residing in the liquid electrolyte. In 1981, Abruna and Bard applied an alternating voltage (with a frequency of 0.5 Hz) to such a modified electrode immersed in a liquid electrolyte solution, so that the organic compound was oxidized during the positive pulses and reduced during the negative pulses.[29] The electrochemically generated ionic radical species thereafter recombined within the solid film to form luminescent excited states, which subsequently released their excess energy via the emission of light. In fact, the entire reaction sequence in hybrid LECs is reminiscent of that of liquid LECs, as outlined in Equations (5.1) through (5.4), with the notable differences being that in hybrid LECs the oxidation and reduction takes place at the same (working) electrode, that the compensating counter-ions initially are residing in a different phase than the electrochemically active organic compound, and that the physical motion of the electrochemically generated ionic radicals (i.e., the oxidized and reduced states of the organic compound) is limited to site-to-site hopping, as convection is excluded in a solid-state environment.

The modified electrode in Reference 29 was fabricated via an *in-situ* electrochemical polymerization of a dissolved Ru-based small molecule compound at a Pt electrode surface, so that a polymer film comprising electroactive Ru(II) centers distributed along long hydrocarbon chains was formed on top of the Pt electrode.[30] In 1994, Richter and coworkers demonstrated an alternative approach to functional modified electrodes for hybrid LECs, when they conveniently deposited the conjugated polymer poly[2-methoxy-5-(2-ethyl-hexyloxy)-1,4-phenylenevinylene] (MEH-PPV) directly from solution on top of a Pt surface.[31] The modified electrode was immersed into an inert liquid electrolyte (in which MEH-PPV is insoluble) in a CV setup, and when the potential applied to the electrode was pulsed between values sufficient to induce electrochemical reduction and oxidation of MEH-PPV, a pumpkin-orange emission was observed.[31] The recorded emission spectrum is characteristic of MEH-PPV, and it was concluded, in agreement with previous studies on hybrid LECs utilizing films of immobilized small molecules for the modified electrode,[29,32–34] that the emission originated in a radiative recombination between electrochemically generated cationic and anionic radicals within the solid conjugated polymer film.[31]

The latter hybrid LECs based on a conjugated polymer-modified electrode are further interesting in the context of electrochemical doping. It is well-established that conjugated polymers can be doped into highly conducting, metallic-like states via electrochemical[35] and chemical methods,[36] and this discovery was in fact rewarded with the Nobel Prize in Chemistry in 2000.[37] The electrochemical doping of a conjugated polymer involves a redox reaction of a polymer segment, which is accompanied by the directed motion of a charge-compensating counter-ion into close proximity of the conjugated polymer chain; the latter being facilitated by the soft nature of the organic film. Specifically, electrochemical p-type doping includes oxidation of the conjugated polymer followed by anionic compensation, while electrochemical n-type doping includes reduction and cationic compensation, in agreement with the reactions outlined in Equations (5.1) and (5.2). It is notable that the issue whether electrochemical doping takes place in solid-state LECs has been the origin of a long-standing debate, and this topic is discussed further in Section 5.4.

5.3 THE SOLID-STATE LEC

In 1995, Pei and coworkers reported the first *solid-state LEC* device.[38] This breakthrough invention benefitted from the discovery and development of solid polymer electrolytes during the preceding two decades,[39,40] as the authors utilized a solid-state blend of a polymer electrolyte and a conjugated polymer (CP) as the active material sandwiched between two electrodes. A typical (and plausibly the most common) device structure of such a "sandwich cell" LEC is presented in Figure 5.1; it

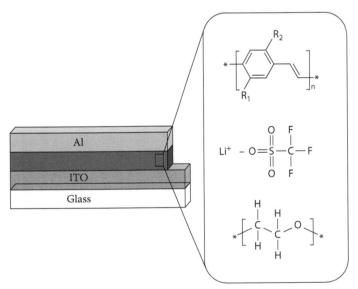

FIGURE 5.1 A solid-state LEC in a conventional sandwich-cell configuration. The inset to the right presents the chemical structures of a typical set of active material components: (from top to bottom) the conjugated polymer MEH-PPV, the salt LiCF$_3$SO$_3$, and the ion-solvation and ion-transporting polymer PEO.

comprises MEH-PPV as the CP, the salt $LiCF_3SO_3$ dissolved in poly(ethylene oxide) (PEO) as the solid-state electrolyte, and Al and indium tin oxide (ITO) as the two charge-injecting electrodes. The entire device structure is mounted on top of a transparent (and mechanically robust) glass substrate so that the light generated within the active material can be efficiently transmitted out of the device through the transparent ITO and glass layers.

The year after, in 1996, Lee and coworkers reported a solid-state LEC device based on a small molecule compound, specifically an ionic transition metal complex (ITMC); in the simplest device architecture, a cationic ruthenium polypyridyl complex with compensating Cl⁻ anions was utilized as the single-component active material sandwiched between Al and ITO electrodes.[41] The first small-molecule-based devices exhibited rather modest brightness and power conversion efficiency, but it was soon demonstrated that significantly better device performance could be attained, for example, via blending the ITMC with a solid electrolyte[42] or via the employment of more appropriate ITMCs, such as $Ru(bpy)^{2+}(X^-)_2$.[43,44]

It is notable that solid-state LEC devices can be very thin, as the thickness of the electrode and active material layers in Figure 5.1 typically are of the order of ~100 nm, and that the total device thickness accordingly in practice is limited by the thickness (and required robustness) of the substrate. Moreover, conformable device configurations are viable via the employment of (transparent) flexible plastic substrates instead of rigid glass.[45,46] In the remainder of this chapter, the primary focus will be on solid-state LEC devices and, in order to distinguish between solid-state LECs based on small molecule compounds and conjugated polymers, from here on we adopt the convention of terming the former *SM-LECs* and the latter *CP-LECs*.

5.4 THE ELUSIVE OPERATIONAL MECHANISM

The conjugated organic compound in the active material of CP-LECs and SM-LECs, as well as in the more commonplace organic light-emitting diodes (OLEDs), can be considered a semiconductor, with an electronic structure characterized by a HOMO, a LUMO, and an energy gap (E_g) that separates the HOMO from the LUMO. (In the case of ordered crystalline organic semiconductors, it is more appropriate to term the HOMO level for the valence band edge and the LUMO level for the conduction band edge.) When the conjugated compound is oxidized or reduced by chemical or electrochemical means, it is convenient, in accordance with semiconductor terminology, to term the process for "doping," the oxidized species for a "hole," and the reduced species for an "electron." These holes and electrons can be mobile and the electronic conductivity (σ) of a doped organic semiconductor is thus given by:

$$\sigma = \sum_{i=n,p} q_i n_i \mu_i$$

(5.5)

where n represents electrons, p represents holes, q_i represents the charge of species i, n_i represents the concentration of species i, and μ_i represents the mobility of species i.

A general feature of solid-state LECs, which clearly distinguishes them from OLEDs, is the existence of mobile ions in the active material. For the CP-LEC device in Figure 5.1, the ion mobility originates in the dissolution of $LiCF_3SO_3$ in PEO to yield "free" Li^+ and $CF_3SO_3^-$ ions; while in SM-LECs based on a single component ITMC such as $Ru(bpy)^{2+}(X^-)_2$, the ionic mobility stems from the dissociation of the comparatively small X^- anions from the large and immobile cationic transition metal complex. In either case, the combined electronic and ionic mobility within the active material plays a critical role in the unique device operation.

Figure 5.2 shows a schematic presentation of the LEC device operation. When a voltage (V) is applied between the two electrodes, the mobile ions redistribute so that thin electric double layers (EDLs) form at the two electrode/active material

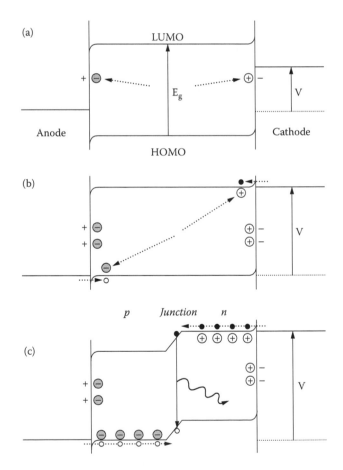

FIGURE 5.2 Schematic electron energy level diagrams for a solid-state LEC comprising a semiconducting organic compound and mobile ions as the active material, following the application of a voltage, V. The large circles represent ions, the small open circles represent holes, and the small solid circles represent electrons. The ionic and electronic responses to V < E_g/e (a) and to V = E_g/e (b) are presented, as well as the steady-state operation at V > E_g/e (c), where the wavy line represents the emission of a photon.

interfaces. At a small applied voltage, this redistribution of ions continues until the bulk of the active material is completely screened from the applied potential, so that the voltage solely drops over the EDLs and no driving electric-field acts on the remaining ionic charge carriers in the bulk of the active material (Figure 5.2a).

At a larger applied voltage, equal to (or exceeding) the so-called "band-gap potential" of the organic semiconductor, $V = E_g/e$, balanced charge injection takes place: electrons are injected at the negative cathode and holes are injected at the positive anode. The injection process is facilitated by the EDLs, since the initial mismatch in energy between the Fermi level of the cathode (anode) and the LUMO (HOMO) level of the organic semiconductor is compensated by the potential drop over the EDLs. The fact that the EDLs are very thin (typically in the sub-nanometer range) allows for highly efficient tunneling of the electronic charger carriers through equally thin energy barriers at the electrode/active material interfaces. The injected electronic charge carriers are thereafter electrostatically compensated by the directed motion of charge-compensating counter ions (Figure 5.2b).

The subsequent processes in solid-state LECs, leading up to the light emission, have been intensely debated for more than a decade. In short, there are two principle models that compete for acceptance: the electrochemical doping model[7,38,47–52] and the electrodynamic model.[10,53–56] The electrochemical doping model states that the injected electronic charge carriers with their compensating counter-ions dope the organic semiconductor, so that its conductivity increases significantly. In this scenario, it is highly plausible that the doping-induced conductivity increase stems from an increase in both the number of charge carriers and their effective mobility [see Equation (5.5)], as elegantly shown and discussed in References 57 and 58. The organic semiconductor is p-type doped at the anode and n-type doped at the cathode, and when these two doped regions make contact in the bulk of the device, a p-n junction is formed. Subsequently injected electronic charge carriers migrate through the highly conducting doped regions, before recombining under the emission of light at the p-n junction (Figure 5.2c). In contrast, the electrodynamic model claims, as stated in the original publication, that "there is little evidence for any binding between individual ions and associated electronic charges, and predictions based on an analogy with chemical doping may be misleading."[54]

In order to distinguish the two models, it is appropriate to measure the electrostatic potential profile in an LEC device during steady-state operation, as the models predict distinctly differing profiles. The electrodynamic model predicts that the entire applied potential will drop over the EDLs at the electrode/active material interfaces, while the electrochemical doping model predicts that a significant fraction of the applied potential will drop over a light-emitting p-n junction.

Figure 5.3 presents such electrostatic potential data recorded on planar "surface cell" devices, comprising a thin film of {MEH-PPV + PEO + KCF$_3$SO$_3$} spin-coated on a glass substrate, on top of which two Au electrodes are deposited.[59] This type of planar surface cell configuration is attractive because the active material is exposed to the environment and as such is prone to direct probing. Figure 5.3a presents an atomic force microscopy topography image, which (together with other data) identifies the electrode/active material interfaces and the 120-μm wide inter-electrode gap, as indicated by the vertical dashed lines. Figure 5.3b and Figure 5.3f present the

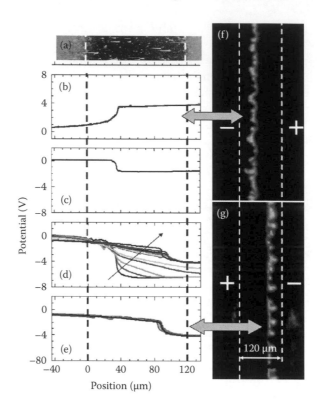

FIGURE 5.3   (a) Two-dimensional topography image of a planar LEC device. (b) Electrostatic potential profile during steady-state operation at V = +5 V. (c) Transient potential profile measured with the device disconnected (open circuit), directly after long-term operation at V = +5 V. (d) Temporal evolution of the potential profile after a subsequent switch to V = –5 V; the arrow indicates increasing time. (e) Steady-state potential profile at V = –5 V. (f) Micrograph showing the light emission during steady-state operation at V = +5 V. (g) Subsequent micrograph showing the light emission from the same device during steady-state operation at V = –5 V. (From P. Matyba, K. Maturova, M. Kemerink, N. D. Robinson, and L. Edman, *Nat Mater* 2009, *8*, 672. With permission.)

electrostatic potential profile (as detected with scanning Kelvin probe microscopy) and the light-emission intensity, respectively, recorded in a device during steady-state operation at $V = 5$ V. It is clear that the steepest potential drop coincides with the position of the distinct light emission zone. Moreover, if the device is disconnected from the voltage supply (i.e., open-circuited), then a (transient) built-in potential is observed at the exact location of the light-emission zone (Figure 5.3c). Accordingly, these data yield strong support for LEC devices operating under the realm of the electrochemical doping model.

At this point, it is appropriate to point out that other data on the electrostatic profile in biased LEC devices are available in the literature,[55,60,61] and in particular that Slinker and coworkers have presented results that are consistent with the electrodynamic model.[55] One key difference between these apparently deviating reports is the choice of organic semiconductor: Matyba and coworkers investigated a CP-LEC system,[59] while Slinker and coworkers studied an SM-LEC system.[55] Consequently, it could very well be that the former system operates in a manner consistent with the electrochemical doping model, while the latter system functions in accordance with the electrodynamic model. It is also important to realize that LEC devices are highly sensitive to electrochemical and chemical side reactions (as discussed in Section 5.6), and that a localized potential drop can stem from the undesired accumulation of insulating side-reactant residues at, for instance, an electrode surface. This risk of misinterpreting data was, however, excluded in the work by Matyba et al. as they also studied their devices under reverse bias, and found that the potential drop and the light-emission zone were positioned at a location that was the exact mirror image of the initial position (i.e., ~35 μm away from the negative cathode; see Figure 5.3e and Figure 5.3g).[59]

An off-centered and reversibly positioned light-emission zone (see Figure 5.3f and Figure 5.3g), in combination with efficient electronic charge injection from essentially any electrode material, is an interesting feature of LECs that can be exploited for the attainment of a bias-direction dependent light-emission color. Yang and Pei fabricated a sandwich-cell device comprising a bi-layer active material, and demonstrated green or orange light emission dependent on the bias direction and in which material the light emission zone was formed.[62] Welter and coworkers reported a similar behavior in LEC-like devices, but their active material was made of a para(phenylene vinylene) (PPV) derivative molecularly doped with a homogeneously dispersed ITMC and the choice of electrode materials was shown to be critically important in order to realize bias-dependent color operation.[63]

Another unique and attractive feature of LECs is that the device operation is not particularly sensitive to the interelectrode gap distance. Gao and coworkers exploited this opportunity to an extreme, when they demonstrated functional planar surface cells with a millimeter-sized interelectrode gap.[64–66] Shin et al. subsequently demonstrated that such wide-gap devices can turn on and emit light at a very low applied voltage close to the band-gap potential,[67,68] and Figure 5.4 presents a set of photographs from the initial operation of an $Au/\{MEH\text{-}PPV+PEO+KCF_3SO_3\}/Au$ surface cell device with a 1-mm interelectrode gap at $V = 5$ V. The photographs were taken in a dark room under UV illumination, since the UV-excited photoluminescence of MEH-PPV (and CPs in general) is sensitively quenched by doping. Accordingly, the dark regions growing from the anodic and cathodic interfaces in Figure 5.4 correspond to doped regions, and when the p- and n-doped regions meet in the bulk of the device, a light-emitting p-n junction is formed, as visualized in the last photograph.

Shin and coworkers utilized the accessible surface cell structure for a relatively accurate calculation of the doping concentration in the doped regions.[69,70] By integrating the current up to the optically detected time for the initial p-n junction formation (using sets of photographs as those presented in Figure 5.4), and dividing this charge with the observed volume of the doped regions, they were able to attain important information regarding the doping structure; notably, they invariably found that the doping concentration in the doped regions is rather high at ~0.1 dopants/CP repeat unit, and that the doping concentration is relatively constant in the entire doped region at the time of the initial p-n junction formation.[69,70] Moreover, the same

FIGURE 5.4   A set of photographs of the initial operation of a planar LEC with a 1-mm inter-electrode gap. The device was operated in a dark room under UV illumination at V = 5 V and T = 360 K. The electrode/active material interfaces are indicated by vertical dashed lines, and the dark regions in the inter-electrode gap correspond to doped areas.

research group was also able to establish that the doping processes in the doped regions can continue after the initial formation of the p-n junction (provided that a surplus of ions exist),[71] and that the doping process will come to a halt under ion-depletion conditions.[72]

5.5   THE TURN-ON TIME

The turn-on time of a light-emitting device is related to the time lag between the application of voltage and the attainment of (significant) light emission. As ions need to redistribute during the initial operation of LEC devices (see Figure 5.2 and related discussion), and the mobility of bulky ions in a solid material typically is low, it is not surprising that the overwhelming consensus in the field is that the turn-on time of conventional solid-state LEC devices to a large degree is dictated by the ionic conductance.

CP-LECs typically contain a blend of a non-polar CP and a polar polymer electrolyte as the active material (see Figure 5.1 for an example). A well-known problem with such non-compatible polymer-based blends is that the two components have a tendency to phase separate from each other. A significant phase separation is undesired from a device perspective because it can lead to problems with one or more of the following key processes: (1) the electronic injection via the EDLs, (2) the ionic screening of the injected electronic charge, and (3) the long-range electronic and ionic transport (the latter being related to the ionic conductance). Microscopy observations of the phase morphology in commonly employed active material blends confirm that the CP and the electrolyte do indeed phase separate, but that the two separate phases can form a bi-continuous morphology, which in principle should allow for reasonable long-range charge transport.[6,26,73] Nevertheless, it is still desirable to minimize or eliminate the phase separation, and a number of successful strategies toward this end include the addition of a surfactant-like additive to the active material,[74] the employment of more compatible CP/electrolyte pairs,[75–80] the grafting of ion-conducting side-chains onto the CP,[81–85] and the employment of a single-component active material in the form of a conjugated polyelectrolyte (CPE) with one of the two ions chemically anchored to the CP backbone.[86–90]

The latter CPEs are similar to the ITMCs used in SM-LECs in that all the required baseline functionalities of an active material in an LEC—ionic and electronic conductivity as well as electroluminescence capacity—are integrated into one single multifunctional compound, which means that phase separation is not an issue. Despite this attractive feature, it is commonly observed that such single-component SM-LECs (and CPE-based LECs) exhibit slower turn-on times than CP-LECs, in many cases on the order of hours, due to a very limited ionic mobility. The latter can, on general terms, be attributed to the facts that the ionic transport in these single-component materials is carried out solely by rather bulky "counter-ions" (e.g., PF_6^- in $Ru(bpy)^{2+}(PF_6^-)_2$) and that these counter-ions are strongly coordinated to very bulky and immobile ionic complexes. Approaches toward larger ionic conductance and faster turn-on time in SM-LECs include the employment of smaller and more mobile counter-ions,[44,91] exposure to humidity or solvent vapors,[92,93] and the blending of an additional electrolyte into the active material.[42,94,95] By using the latter strategy, Slinker and coworkers achieved turn-on times of the order of 1 to 10 ms, and used this notable feat to demonstrate a cascaded device structure that emitted constant light to the eye when biased directly from a standard U.S. outlet (110-120 V RMS, 60 Hz).[94]

A radically different approach toward *extremely* fast-response operation of solid-state LEC is to "permanently" stabilize a desired ionic (and doping) distribution, so that the subsequent operation is de-coupled from slow ionic motion. Gao and coworkers turned-on a CP-LEC in the normal manner via ionic redistribution (see Figure 5.2), and then cooled the device under applied bias to a low temperature ($T <$ 200 K) at which the ionic mobility is effectively zero.[96,97] With this so-called "frozen junction" approach, the authors were able to demonstrate a very fast turn-on time at the low temperature of the order of ~1 μs for a sandwich-cell CP-LEC.[97] Subsequent efforts have demonstrated that active materials with effectively zero ionic mobility at room temperature (RT), but reasonable ionic mobility at elevated temperatures, can be utilized for a more practical RT "frozen junction" operation.[79,98,99] Leger et al. introduced a novel and promising concept for permanently stabilized doping profiles, and a related fast turn-on time, when they utilized polymerizable ions as the dopant counter-ions and managed to chemically lock the desired ionic distribution into place, presumably via an ionic network formation.[100] Another notably novel means toward stabilized junction formation and fast device turn-on constitutes the fabrication of an active material comprising a cationic and anionic CPE bilayer structure, and the subsequent solvent-induced removal of the mobile counter-ions.[101] The topic of permanently stabilized doping profiles in LEC devices is discussed in more depth and detail in Chapter 9.

5.6 TOWARD EFFICIENT AND LONG-TERM OPERATION

Two important figure-of-merits for a light-emitting device from an application viewpoint are the operational lifetime, and how efficient input electric energy is transformed into visible light as quantified by the power conversion efficacy (PCE). The conventional unit for visible light power is lumen (lm), which takes into account that our eyes exhibit a strongly wavelength-dependent sensitivity, with the highest sensitivity in the green region at a wavelength of 555 nm. Accordingly, the maximum PCE value for such a monochromatic green-emitting device is 683 lm/W, while an "ideal" white-emitting device exhibits a lower value of ~240 lm and red- and blue-emitting devices demonstrate even lower maximum PCE values. As a further point of reference, it is of interest to note that the common (but soon to be phased out) light bulb has a PCE value of ~10 lm/W and an operational lifetime of ~1000 h, while its expected short-term replacement the compact fluorescent lamp (commonly termed "the energy saving lamp") typically exhibits a PCE value of ~60 lm/W and an operational lifetime of ~10.000 h.

Already at the time of their initial demonstration, it was realized that "optimized" solid-state LEC devices have the potential for very efficient operation and high PCE values. The unique operational mechanism allows for balanced injection and efficient transport of electrons and holes at a low voltage close to the band-gap potential. Moreover, in devices with the emission zone positioned in the middle of the active material, far away from the electrodes (see Figure 5.4), the recombination of electrons and holes into electronically excited states (excitons) will be very high, while the electrode-induced non-radiative quenching of excitons is minimized. The final conversion of excitons into light emission within the emission zone (the p-n junction) and the out-coupling of the generated light out of the device are, however, issues that are in need of further studies. Direct observations of CP-based surface cells during operation reveal that the light emission originates in very close proximity to highly doped regions with quenched luminescence, which indicates that doping-induced quenching of excitons might be a serious loss mechanism in LEC devices. This hypothesis is supported by recent studies, where a significant improvement in the efficiency could be attained via a controlled expansion of the light-emission zone.[6,102,103] Moreover, self-quenching in general has been identified as a significant problem in solid-state LEC devices, and successful efforts toward alleviating this problem include the addition of a diluting inert polymer to the active layer,[75,91,104] the introduction of bulky side chains on the ligands of the emissive ITMC complex in SM-LECs,[91,105,106] and energy transfer within a host-guest system.[107]

When reviewing the reported efficiency values of different LEC devices, a few words of caution are in place: the measurement of exact PCE values requires relatively sophisticated and accurately calibrated set-ups, and as the performance of LEC devices approaches commercial viability, more attention needs to be directed toward the implementation of such appropriate measurement procedures.[108] In addition, for applications, it is not necessarily the maximum PCE value per se that is of interest, but rather the PCE value at significant brightness. With that said, we turn to the literature. For CP-LECs, the top PCE values reported up to date are in the range of between 1 and 12 lm/W,[45,74,75,85,109–112] with the highest value (12 lm/W) attained from a blue-green emitting polyfluorene-based device by Yang and Pei in 1997.[85] SM-LEC devices exhibit a distinct advantage over CP-LECs in that the emissive ITMC can harvest both triplet and singlet excitons for light emission, while solely singlet excitons are emissive in CPs. Thus, based on spin statistics, SM-LECs are expected to have a four times higher efficiency than CP-LECs. The record efficiency for SM-LECs has seen a steady increase since 1997, as detailed in Reference 10, and the best devices as-of-today exhibit PCE values of ~6 lm/W for Ru-based devices,[91,113] ~25 to 35 lm/W for Ir-based devices,[106,107,114,115] and ~30 lm/W for recently demonstrated Cu-based SM-LEC devices.[116] Finally, it is notable that white-emitting LEC devices, of particular interest for solid-state lighting applications, have been reported utilizing CPs[85] and SMs[117] as the active material.

We now shift our attention to plausibly the largest obstacle for a large-scale introduction of LEC devices in applications: the limited operational lifetime. It has become clear that the unique and attractive operational mechanism comes with a price; namely, that LEC devices are sensitive to a large plethora of undesired chemical and electrochemical "side reactions" that can potentially severely limit their stability and performance. A number of different side reaction mechanisms have been proposed and verified,[26,45,118,119] and direct evidence has, in some cases, even been provided via the identification of side reaction residues.[69,105,120,121] Of particular interest are those reports in which the identification of a side reaction mechanism is accompanied by a recipe for the alleviation of its impact and the demonstration of devices with improved operational lifetime; a few particularly notable studies are presented later. First another word of caution: the term *operational lifetime* has been defined differently in different research groups, and a long lifetime at low brightness might not necessarily be preferable over a shorter lifetime at higher brightness. In addition, some of the reported data stem from extrapolations, which might be inappropriate because the final decay of an LEC device can be rather abrupt, as visualized in Figure 5.5.

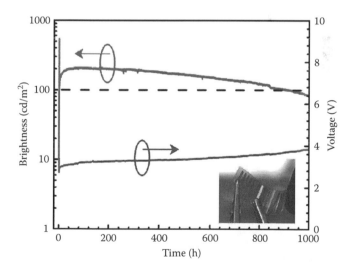

FIGURE 5.5 The temporal evolution of the brightness and the voltage of an ITO/PEDOT-PSS/{MEH-PPV:PEO:KCF$_3$SO$_3$}/Al sandwich cell with an active material mass ratio of 1:0.085:0.03. The device was operated at T = 298 K and in galvanostatic mode. The initial "pre-bias" current, $I_{pre-bias}$ = 0.005 A, was applied for t = 0.5 h, and it was followed by long-term operation at I = 0.001 A. The inset presents a flexible device during operation at T = 298 K and I = 0.005 A. (From J. Fang, P. Matyba, L. Edman, *Advanced Functional Materials* 2009, *19*, 2671. With permission.)

Cao et al. brought forward the hypothesis that significant, and with time increasing, phase separation between a hydrophobic CP and a hydrophilic electrolyte in the active material is detrimental for the long-term stability of CP-LECs. These authors introduced two functional approaches that addressed this problem and resulted in devices with an operational lifetime of ~100 h at high brightness.[74,76] Rudmann and coworkers demonstrated that blending an inert and glassy polymer (such as PMMA) with the ITMC in SM-LECs can improve the device stability and reliability.[104] Shao et al. thereafter employed a combination of these concepts—the utilization of a compatible (hydrophobic) CP/electrolyte pair and the blending of an inert polymer into the active material—and attained a respectable operational lifetime of several hundred hours for a yellow-emitting CP-LEC.[75]

It has been demonstrated further that the electrochemical stability window of the electrode materials and the electrolyte plays a critical role for the device stability, particularly how it is positioned in comparison to the electrochemical window spanned by the reduction and oxidation levels of the electrochemically active organic compound.[26,69,122,123] For instance, it was recently shown that the electrolyte in a commonplace CP-LEC (with similar active material constituents as specified Figure 5.1) exhibits a reduction level that is positioned at a lower potential than the n-type doping level of the CP, but that this propensity for an undesired electrochemical side-reaction at the cathodic interface can be alleviated by operating the device at a high initial voltage.[69] Moreover, postmortem studies of open planar surface cells with the same active material have revealed that the ultimate degradation reaction takes place at the location of the emission zone (the p-n junction), presumably via interactions between excitons and the electrolyte constituents.[121] With this information at hand, Fang and coworkers were recently able to design and realize orange-red emitting CP-LECs with an operational lifetime of ~1000 h at significant brightness, as presented in Figure 5.5.[45] Moreover, the same authors were able to attain a similar stabile and efficient performance for flexible devices fabricated on plastic PET substrates, and one such flexed device during operation is displayed in the inset of Figure 5.5.

Finally, it is relevant to emphasize that the preparation and characterization procedures of LEC devices are extremely important, as it is, for example, well established that small traces of water, oxygen, and solvents within the active material can have a profound influence on the device performance.[92,120,124] Moreover, LEC devices based on nominally identical configurations have been reported with drastically deviating performance, presumably due to a varying degree of quality and purity of the active material constituents and inappropriate procedures during device fabrication and testing. Thus, to state it bluntly, the devil is in the details when it comes to the appropriate evaluation of LEC performance.

5.7 CONCLUDING REMARKS

The solid-state LEC simultaneously represents an opportunity for exciting and novel science and the promise for important and ubiquitous future applications. The characteristic mixed electron- and ion-transport capacity of the active material constitutes the basis for the complex and intensely debated operational mechanism, but also allows for a wide range of very attractive device properties. The fact that the efficient operation of LEC devices, in contrast with competing emerging light-emission technologies, to a large degree is independent of the work function of the electrode materials and the thickness of the active layer is important because it promises to allow for an unprecedented low-cost and fault-tolerant fabrication of flexible, lightweight and large-area light-emitting devices with roll-to-roll printing technologies.

The unique operational mechanism that includes an initial in-situ redistribution of ionic carriers comes with a price in a relatively slow turn-on time for light emission, which in turn excludes the possibility of utilizing conventional LEC devices in fast video-quality displays. However, a number of recent novel approaches aimed at a permanent stabilization of a desired ion distribution are interesting in this context, as they might pave the way for ultra-fast LEC operation. The efficiency of state-of-the-art LECs is reasonably high and, particularly devices based on ionic transition metal complexes are approaching the so-called "energy-saving lamps" in how efficient they transform input electric energy to visible light. Further improvement in this area can still be expected, particularly via an improved understanding and optimization of the processes related to the conversion of excitons into light exiting the device.

The critical problem with the current generation of LEC devices is an inadequate operational lifetime; but also recent reports, based on an improvement in material properties and the understanding of the governing device physics, indicate promise, as significant operational lifetimes of the order of 1000 h at high brightness have been attained. To summarize, in consideration of the complex, and not yet fully understood, operation of LECs and the well-documented importance of high-quality materials and procedures, it is wise if basic science can continue to accompany more technologically inclined activities within this vibrant field so that the LEC can realize its true bright potential.

5.8   ACKNOWLEDGMENTS

The author is grateful to past and present members of the Organic Photonics and Electronics Group at Umeå University, Sweden, and other collaborators around the world for stimulating and challenging discussions over the years. The author is a Royal Swedish Academy of Sciences Research Fellow supported by a grant from the Knut and Alice Wallenberg Foundation.

REFERENCES

1. D. M. Hercules, *Science* 1964, *145*, 808.
2. L. B. Anderson, C. N. Reilley, *J. Electroanal. Chem.* 1965, *10*, 538.
3. Chandros.Ea, Longwort.Jw, R. E. Visco, *J. Am. Chem. Soc.* 1965, *87*, 3259.
4. L. R. Faulkner, A. J. Bard, *J. Am. Chem. Soc.* 1968, *90*, 6284.
5. N. R. Armstrong, R. M. Wightman, E. M. Gross, *Light-emitting electrochemical processes*, Vol. 52, 2001.
6. Q. J. Sun, Y. F. Li, Q. B. Pei, *J. Disp. Technol.* 2007, *3*, 211.
7. L. Edman, *Electrochimica Acta* 2005, *50*, 3878.
8. A. J. Bard, Ed. *Electrogenerated chemiluminescence*, Marcel Dekker, New York 2004.
9. J. M. Leger, *Adv. Mater.* 2008, *20*, 837.

10. J. D. Slinker, J. Rivnay, J. S. Moskowitz, J. B. Parker, S. Bernhard, H. D. Abruna, G. G. Malliaras, *Journal of Materials Chemistry* 2007, *17*, 2976.
11. D. Dini, *Chem. Mat.* 2005, *17*, 1933.
12. E. Holder, B. M. W. Langeveld, U. S. Schubert, *Adv. Mater.* 2005, *17*, 1109.
13. www.add-vision.com.
14. S. A. Cruser, A. J. Bard, *J. Am. Chem. Soc.* 1969, *91*, 267.
15. H. Schaper, H. Kostlin, E. Schnedler, *J. Electrochem. Soc.* 1982, *129*, 1289.
16. M. Orlik, K. Doblhofer, G. Ertl, *J. Phys. Chem. B* 1998, *102*, 6367.
17. D. Laser, A. J. Bard, *J. Electrochem. Soc.* 1975, *122*, 632.
18. J. S. Dunnett, M. Voinov, *J. Electroanal. Chem.* 1978, *89*, 181.
19. J. S. Dunnett, M. Voinov, *Journal of the Chemical Society-Faraday Transactions I* 1977, *73*, 853.
20. N. E. Tokeltak, R. E. Hemingwa, A. J. Bard, *J. Am. Chem. Soc.* 1973, *95*, 6582.
21. D. J. Vinyard, S. J. Su, M. M. Richter, *J. Phys. Chem. A* 2008, *112*, 8529.
22. S. C. Chang, Y. Yang, Q. B. Pei, *Appl. Phys. Lett.* 1999, *74*, 2081.
23. S. C. Chang, Y. Yang, *Appl. Phys. Lett.* 1999, *75*, 2713.
24. S. C. Chang, Y. F. Li, Y. Yang, *J. Phys. Chem. B* 2000, *104*, 11650.
25. J. B. Edel, A. J. deMello, J. C. deMello, *Chem. Commun.* 2002, 1954.
26. P. Matyba, M. R. Andersson, L. Edman, *Organic Electronics* 2008, *9*, 699.
27. Y. F. Li, Y. Cao, J. Gao, D. L. Wang, G. Yu, A. J. Heeger, *Synth. Met.* 1999, *99*, 243.
28. Q. J. Sun, H. Q. Wang, C. H. Yang, G. F. He, Y. F. Li, *Synth. Met.* 2002, *128*, 161.
29. H. D. Abruna, A. J. Bard, *J. Am. Chem. Soc.* 1982, *104*, 2641.
30. H. D. Abruna, P. Denisevich, M. Umana, T. J. Meyer, R. W. Murray, *J. Am. Chem. Soc.* 1981, *103*, 1.
31. M. M. Richter, F. R. F. Fan, F. Klavetter, A. J. Heeger, A. J. Bard, *Chem. Phys. Lett.* 1994, *226*, 115.
32. I. Rubinstein, A. J. Bard, *J. Am. Chem. Soc.* 1980, *102*, 6641.
33. I. Rubinstein, A. J. Bard, *J. Am. Chem. Soc.* 1981, *103*, 5007.
34. F. R. F. Fan, A. Mau, A. J. Bard, *Chem. Phys. Lett.* 1985, *116*, 400.
35. P. J. Nigrey, A. G. Macdiarmid, A. J. Heeger, *J. Chem. Soc.-Chem. Commun.* 1979, 594.
36. H. Shirakawa, E. J. Louis, A. G. Macdiarmid, C. K. Chiang, A. J. Heeger, *J. Chem. Soc.-Chem. Commun.* 1977, 578.
37. http://nobelprize.org/nobel_prizes/chemistry/laureates/2000/public.html.
38. Q. B. Pei, G. Yu, C. Zhang, Y. Yang, A. J. Heeger, *Science* 1995, *269*, 1086.
39. D. E. Fenton, J. M. Parker, P. V. Wright, *Polymer* 1973, *14*, 589.
40. P. G. Bruce, Ed. *Solid State Electrochemistry*, Cambridge University Press, 1995.
41. J. K. Lee, D. S. Yoo, E. S. Handy, M. F. Rubner, *Appl. Phys. Lett.* 1996, *69*, 1686.
42. C. H. Lyons, E. D. Abbas, J. K. Lee, M. F. Rubner, *J. Am. Chem. Soc.* 1998, *120*, 12100.
43. E. S. Handy, A. J. Pal, M. F. Rubner, *J. Am. Chem. Soc.* 1999, *121*, 3525.
44. F. G. Gao, A. J. Bard, *J. Am. Chem. Soc.* 2000, *122*, 7426.
45. J. Fang, P. Matyba, L. Edman, *Advanced Functional Materials* 2009, *19*, 2671.
46. G. Santos, F. J. Fonseca, A. M. Andrade, A. O. T. Patrocinio, S. K. Mizoguchi, N. Y. M. Iha, M. Peres, W. Simoes, T. Monteiro, L. Pereira, "Opto-electrical properties of single layer flexible electroluminescence device with ruthenium complex," presented at *22nd International Conference on Amorphous and Nanocrystalline Semiconductors*, Breckenridge, CO, Aug 19–24, 2007.
47. Q. Pei, A. J. Heeger, *Nat. Mater.* 2008, *7*, 167.
48. Q. B. Pei, Y. Yang, G. Yu, C. Zhang, A. J. Heeger, *J. Am. Chem. Soc.* 1996, *118*, 3922.
49. N. D. Robinson, J. H. Shin, M. Berggren, L. Edman, *Phys Rev B* 2006, *74*, 155210.
50. D. J. Dick, A. J. Heeger, Y. Yang, Q. B. Pei, *Adv. Mater.* 1996, *8*, 985.
51. H. Rudmann, S. Shimada, M. F. Rubner, *J. Appl. Phys.* 2003, *94*, 115.
52. D. T. Simon, D. B. Stanislowski, S. A. Carter, *Appl. Phys. Lett.* 2007, *90*.

53. J. C. deMello, *Phys Rev B* 2002, *66*, 235210.
54. J. C. deMello, N. Tessler, S. C. Graham, R. H. Friend, *Phys Rev B* 1998, *57*, 12951.
55. J. D. Slinker, J. A. DeFranco, M. J. Jaquith, W. R. Silveira, Y. W. Zhong, J. M. Moran-Mirabal, H. G. Craighead, H. D. Abruna, J. A. Marohn, G. G. Malliaras, *Nat. Mater.* 2007, *6*, 894.
56. G. G. Malliaras, J. D. Slinker, J. A. DeFranco, M. J. Jaquith, W. R. Silveira, Y. W. Zhong, J. M. Moran-Mirabal, H. G. Craighead, H. D. Abruna, J. A. Marohn, *Nat. Mater.* 2008, *7*, 168.
57. V. I. Arkhipov, E. V. Emelianova, P. Heremans, H. Bassler, *Phys. Rev. B* 2005, *72*, 5.
58. H. Shimotani, G. Diguet, Y. Iwasa, *Appl. Phys. Lett.* 2005, *86*, 3.
59. P. Matyba, K. Maturova, M. Kemerink, N. D. Robinson, L. Edman, *Nat Mater* 2009, *8*, 672.
60. L. S. C. Pingree, D. B. Rodovsky, D. C. Coffey, G. P. Bartholomew, D. S. Ginger, *J. Am. Chem. Soc.* 2007, *129*, 15903.
61. L. S. C. Pingree, O. G. Reid, D. S. Ginger, *Adv. Mater.* 2009, *21*, 19.
62. Y. Yang, Q. B. Pei, *Appl. Phys. Lett.* 1996, *68*, 2708.
63. S. Welter, K. Brunner, J. W. Hofstraat, L. De Cola, *Nature* 2003, *421*, 54.
64. J. Gao, J. Dane, *Appl. Phys. Lett.* 2003, *83*, 3027.
65. J. Gao, J. Dane, *Appl. Phys. Lett.* 2004, *84*, 2778.
66. J. Gao, J. Dane, *J. Appl. Phys.* 2005, *98*.
67. J. H. Shin, A. Dzwilewski, A. Iwasiewicz, S. Xiao, A. Fransson, G. N. Ankah, L. Edman, *Appl. Phys. Lett.* 2006, *89*.
68. J. H. Shin, L. Edman, *J. Am. Chem. Soc.* 2006, *128*, 15568.
69. J. Fang, P. Matyba, N. D. Robinson, L. Edman, *J. Am. Chem. Soc.* 2008, *130*, 4562.
70. J. H. Shin, N. D. Robinson, S. Xiao, L. Edman, *Advanced Functional Materials* 2007, *17*, 1807.
71. N. D. Robinson, J. F. Fang, P. Matyba, L. Edman, *Phys. Rev. B* 2008, *78*, 7.
72. J. F. Fang, Y. L. Yang, L. Edman, *Appl. Phys. Lett.* 2008, *93*, 3.
73. M. A. Summers, S. K. Buratto, L. Edman, *Thin Solid Films* 2007, *515*, 8412.
74. Y. Cao, G. Yu, A. J. Heeger, C. Y. Yang, *Appl. Phys. Lett.* 1996, *68*, 3218.
75. Y. Shao, G. C. Bazan, A. J. Heeger, *Adv. Mater.* 2007, *19*, 365.
76. Y. Cao, Q. B. Pei, M. R. Andersson, G. Yu, A. J. Heeger, *J. Electrochem. Soc.* 1997, *144*, L317.
77. J. S. Wilson, M. J. Frampton, J. J. Michels, L. Sardone, G. Marletta, R. H. Friend, P. Samori, H. L. Anderson, F. Cacialli, *Adv. Mater.* 2005, *17*, 2659.
78. F. P. Wenzl, P. Polt, A. Haase, S. Patil, U. Scherf, G. Leising, "Microstructure tailoring of conjugated polymer-electrolyte blends for light-emitting electrochemical cells," presented at *7th International Symposium on Systems with Fast Ionic Transport (ISSFIT)*, Bled, SLOVENIA, May 05-09, 2004.
79. C. H. Yang, Q. J. Sun, J. Qiao, Y. F. Li, *J. Phys. Chem. B* 2003, *107*, 12981.
80. T. Ouisse, M. Armand, Y. Kervella, O. Stephan, *Appl. Phys. Lett.* 2002, *81*, 3131.
81. L. Zhu, X. Z. Tang, *J. Appl. Polym. Sci.* 2007, *104*, 1118.
82. F. P. Wenzl, A. Fian, P. Polt, A. Rudorfer, G. Leising, *Electrochim. Acta* 2007, *52*, 6229.
83. J. Morgado, R. H. Friend, F. Cacialli, B. S. Chuah, S. C. Moratti, A. B. Holmes, *J. Appl. Phys.* 1999, *86*, 6392.
84. S. Tasch, L. Holzer, F. P. Wenzl, J. Gao, B. Winkler, L. Dai, A. W. H. Mau, R. Sotgiu, M. Sampietro, U. Scherf, K. Mullen, A. J. Heeger, G. Leising, "Light-emitting electrochemical cells with microsecond response times based on PPPs and novel PPVs," presented at *International Conference on Science and Technology of Synthetic Metals (ICSM 98)*, Montpellier, France, Jul 12–18, 1998.
85. Y. Yang, Q. B. Pei, *J. Appl. Phys.* 1997, *81*, 3294.

86. V. Cimrova, W. Schmidt, R. Rulkens, M. Schulze, W. Meyer, D. Neher, *Adv. Mater.* 1996, *8*, 585.

87. L. Edman, B. Liu, M. Vehse, J. Swensen, G. C. Bazan, A. J. Heeger, *J. Appl. Phys.* 2005, *98*.

88. L. Edman, M. Pauchard, B. Liu, G. Bazan, D. Moses, A. J. Heeger, *Appl. Phys. Lett.* 2003, *82*, 3961.

89. D. Vak, S. H. Oh, D. Y. Kim, *Appl. Phys. Lett.* 2009, *94*, 3.

90. C. V. Hoven, A. Garcia, G. C. Bazan, T. Q. Nguyen, *Adv. Mater.* 2008, *20*, 3793.

91. H. Rudmann, S. Shimada, M. F. Rubner, *J. Am. Chem. Soc.* 2002, *124*, 4918.

92. D. L. Pile, A. J. Bard, *Chem. Mat.* 2005, *17*, 4212.

93. W. Zhao, C. Y. Liu, Q. Wang, J. M. White, A. J. Bard, *Chem. Mat.* 2005, *17*, 6403.

94. J. D. Slinker, J. Rivnay, J. A. DeFranco, D. A. Bernards, A. A. Gorodetsky, S. T. Parker, M. P. Cox, R. Rohl, G. G. Malliaras, S. Flores-Torres, H. D. Abruna, *J. Appl. Phys.* 2006, *99*, 5.

95. S. T. Parker, J. D. Slinker, M. S. Lowry, M. P. Cox, S. Bernhard, G. G. Malliaras, *Chem. Mat.* 2005, *17*, 3187.

96. J. Gao, G. Yu, A. J. Heeger, *Appl. Phys. Lett.* 1997, *71*, 1293.

97. J. Gao, Y. F. Li, G. Yu, A. J. Heeger, *J. Appl. Phys.* 1999, *86*, 4594.

98. G. Yu, Y. Cao, M. Andersson, J. Gao, A. J. Heeger, *Adv. Mater.* 1998, *10*, 385.

99. L. Edman, M. Pauchard, D. Moses, A. J. Heeger, *J. Appl. Phys.* 2004, *95*, 4357.

100. J. M. Leger, D. B. Rodovsky, G. R. Bartholomew, *Adv. Mater.* 2006, *18*, 3130.

101. C. H. W. Cheng, M. C. Lonergan, *J. Am. Chem. Soc.* 2004, *126*, 10536.

102. Y. G. Zhang, Y. F. Hu, J. Gao, *Appl. Phys. Lett.* 2006, *88*.

103. P. Pachler, F. P. Wenzl, U. Scherf, G. Leising, *J. Phys. Chem. B* 2005, *109*, 6020.

104. H. Rudmann, M. F. Rubner, *J. Appl. Phys.* 2001, *90*, 4338.

105. S. Bernhard, J. A. Barron, P. L. Houston, H. D. Abruna, J. L. Ruglovksy, X. C. Gao, G. G. Malliaras, *J. Am. Chem. Soc.* 2002, *124*, 13624.

106. H. J. Bolink, E. Coronado, R. D. Costa, N. Lardies, E. Orti, *Inorg. Chem.* 2008, *47*, 9149.

107. H. C. Su, C. C. Wu, F. C. Fang, K. T. Wong, *Appl. Phys. Lett.* 2006, *89*, 3.

108. S. R. Forrest, D. D. C. Bradley, M. E. Thompson, *Adv. Mater.* 2003, *15*, 1043.

109. H. Q. Wang, Q. J. Sun, Y. F. Li, L. Duan, Y. Qiu, X. Y. Li, *Polym. Int.* 2003, *52*, 343.

110. L. Edman, D. Moses, A. J. Heeger, *Synth. Met.* 2003, *138*, 441.

111. Y. Shao, X. Gong, A. J. Heeger, M. Liu, A. K. Y. Jen, *Adv. Mater.* 2009, *21*, 1972.

112. Z. B. Yu, M. L. Sun, Q. B. Pei, *J. Phys. Chem. B* 2009, *113*, 8481.

113. C. Y. Liu, A. J. Bard, *Appl. Phys. Lett.* 2005, *87*, 3.

114. H. C. Su, F. C. Fang, T. Y. Hwu, H. H. Hsieh, H. F. Chen, G. H. Lee, S. M. Peng, K. T. Wong, C. C. Wu, *Advanced Functional Materials* 2007, *17*, 1019.

115. A. B. Tamayo, S. Garon, T. Sajoto, P. I. Djurovich, I. M. Tsyba, R. Bau, M. E. Thompson, *Inorg. Chem.* 2005, *44*, 8723.

116. Q. S. Zhang, Q. G. Zhou, Y. X. Cheng, L. X. Wang, D. G. Ma, X. B. Jing, F. S. Wang, *Advanced Functional Materials* 2006, *16*, 1203.

117. H. C. Su, H. F. Chen, F. C. Fang, C. C. Liu, C. C. Wu, K. T. Wong, Y. H. Liu, S. M. Peng, *J. Am. Chem. Soc.* 2008, *130*, 3413.

118. Y. Kervella, M. Armand, O. Stephan, *J. Electrochem. Soc.* 2001, *148*, H155.

119. J. Dane, J. Gao, *Appl. Phys. Lett.* 2004, *85*, 3905.

120. G. Kalyuzhny, M. Buda, J. McNeill, P. Barbara, A. J. Bard, *J. Am. Chem. Soc.* 2003, *125*, 6272.

121. T. Wågberg, P. R. Hania, N. D. Robinson, J. H. Shin, P. Matyba, L. Edman, *Adv. Mater.* 2008, *20*, 1744.

122. J. H. Shin, P. Matyba, N. D. Robinson, L. Edman, *Electrochim. Acta* 2007, *52*, 6456.

123. T. Johansson, W. Mammo, M. R. Andersson, O. Inganas, *Chem. Mat.* 1999, *11*, 3133.

124. M. Buda, G. Kalyuzhny, A. J. Bard, *J. Am. Chem. Soc.* 2002, *124*, 6090.

6 Fixed Junction Light-Emitting Electrochemical Cells

Janelle Leger and Amanda Norell Bader

CONTENTS

6.1 INTRODUCTION

In 1995, Pei et al. demonstrated a device in which ion incorporation in a polymer light-emitting diode architecture leads to dynamic, symmetric light emission in a device with reduced thickness dependence and reduced dependence on electrode work function.[1] The so-called polymer light-emitting electrochemical cell (LEC) introduced a new approach for the operation of a solution-processed optoelectronic device. In essence, the LEC was first described as a dynamic, self-assembled polymer *p-i-n* junction in which *p*- and *n*-type doping occur via electrochemical oxidation and reduction, respectively.[2] In an LEC, the electric field and light emission occur over a thin intrinsic (non-doped) layer often positioned close to the cathode (Figure 6.1). Several models have been described to explain the mechanism of operation of these devices.[1–6] It is generally agreed, however, that in an LEC, the cations and anions dissociate under an applied electric field and accumulate along opposite electrode/polymer interfaces. Device characteristics can be impacted both through induced electrochemical doping and through the buildup of uncompensated charges at the electrodes. The increased conductivity of the doped material reduces barriers to charge injection and allows the potential in the device to drop over a relatively thin undoped region. Ions not ultimately compensated by doped polymer contribute to these effects by introducing a dipole layer, which results in a potential drop at the electrode interface. The history, mechanism, and progress of LEC operation and current progress in the field are the subjects of Chapter 5.

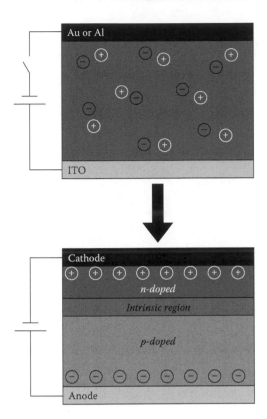

FIGURE 6.1 Schematic depicting the generalized understanding of the operation of a polymer light-emitting electrochemical cell. The anions and cations from the salt move when subjected to an applied bias, then act as counterions in the electrochemical doping of the light-emitting polymer. Emission occurs over a thin, non-doped, or intrinsic, region in the film.

One of the key defining features of the LEC, in contrast with the traditional polymer light-emitting diode (LED), is its reversible nature.[7] Because ion migration and electrochemical doping are ultimately reversible processes, a change in the applied bias will result in a change in the ion distribution throughout the polymer film. Indeed, switching the direction of the applied bias will result in the establishment of a nearly mirror distribution of counterions within the device along the opposite electrodes. There are potential advantages to having a dynamic junction if switching speeds are high enough. For example, several groups have reported the ability to achieve two-color emission from a single LEC by switching the bias orientation with respect to a two-layer film containing different emissive species.[2,8] However, in general, the dynamic nature of the ion distribution in an LEC is undesirable, as it results in a number of problems related to stability and versatility. Therefore, in an effort to improve on the performance and application potential of the LEC, considerable efforts have been directed at creating a fixed-junction LEC configuration, or a method for immobilizing the counterion and electrochemical doping distribution in a polymer film, in order to allow the in situ construction of a stable polymer p-i-n junction (Figure 6.2).

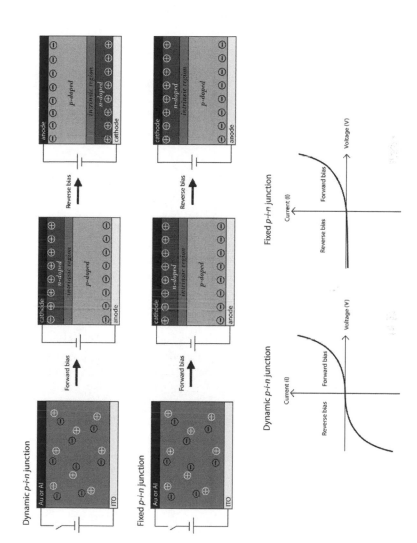

FIGURE 6.2 Fixed vs. dynamic LEC behavior and properties. In a fixed LEC, the counterions no longer move under an applied bias once immobilized. The resulting device displays rectification, which is displayed as a non-symmetric current-voltage curve.

Several methods for achieving such a fixed junction have been demonstrated to date; the earliest reports took advantage of the temperature-dependent ionic mobility in the solid-state film. After a desired ion distribution is obtained in this system, the temperature of the device is reduced to a point where the ions are no longer mobile, thereby preventing them from redistributing under a change in applied voltage. Other efforts focus on taking advantage of chemical bonding techniques to immobilize counterions in situ. For example, building ionic groups into the polymer structure itself, building polymerizable groups onto an ionic species followed by counterions via in situ covalent bond formation, or using self-assembled monolayers with charged head groups at opposite electrodes are all methods that have been demonstrated and can potentially provide a simple method for creating a fixed polymer junction that is compatible with large-scale processing techniques. The detailed motivations for the creation of a fixed junction LEC, the process and performance of the methods demonstrated thus far for achieving a fixed junction, and the remaining challenges of this field are the subject of this chapter.

6.2   MOTIVATIONS FOR CREATING A FIXED JUNCTION LEC

As noted previously, the LEC is a dynamic structure, meaning that the profile of counterions throughout the device depends on the magnitude and direction of the applied bias. The dynamic nature of the traditional LEC causes a number of issues with device operation that can potentially be improved by finding a way to precisely control the distribution of ionic carriers within the device. Because ions redistribute to a neutral distribution when the applied bias is removed, the turn-on time of an LEC is limited primarily by the low ionic mobility in the polymer films.[9,10] The traditionally slow turn-on time seen in dynamic junction LECs is one of the limiting factors in the application of LECs to commercially viable display technologies. While turn-on times on the order of milliseconds are required for modern display applications, the turn-on times of dynamic LECs range from seconds to minutes in most systems.[9,10] In principle, establishing a counterion distribution using an applied bias, followed by ion immobilization, will allow the desired distribution to remain unchanged despite changes to the magnitude or direction of the applied bias. Therefore, this arrangement should allow for near instantaneous turn-on for these devices. To the extent that the turn-on times have been systematically studied in LECs, improvements have been demonstrated, with fixed junction LECs showing an order of magnitude improvement in turn-on response over their dynamic LEC counterparts.[7,11,12]

Another performance-limiting factor related to the dynamic nature of traditional LECs is lifetime.[10,13–15] The relatively short lifetime of LECs is thought to be related to certain electrochemical reactions that occur in the presence of high voltage or high concentrations of counterions.[16–18] As it is currently understood, the electrochemical doping reaction in these conjugated polymers introduces species known as polarons and bipolarons through a rearrangement of the bonding character along the polymer backbone (Figure 6.3).[19,20] This rearrangement results in a shift, for example, from benziodal to quinoidal forms, in the conjugated rings. These electrochemical processes are discussed in more detail in Chapter 1. The conjugated polymers used as the active emissive species in LECs are also known to undergo overdoping, which

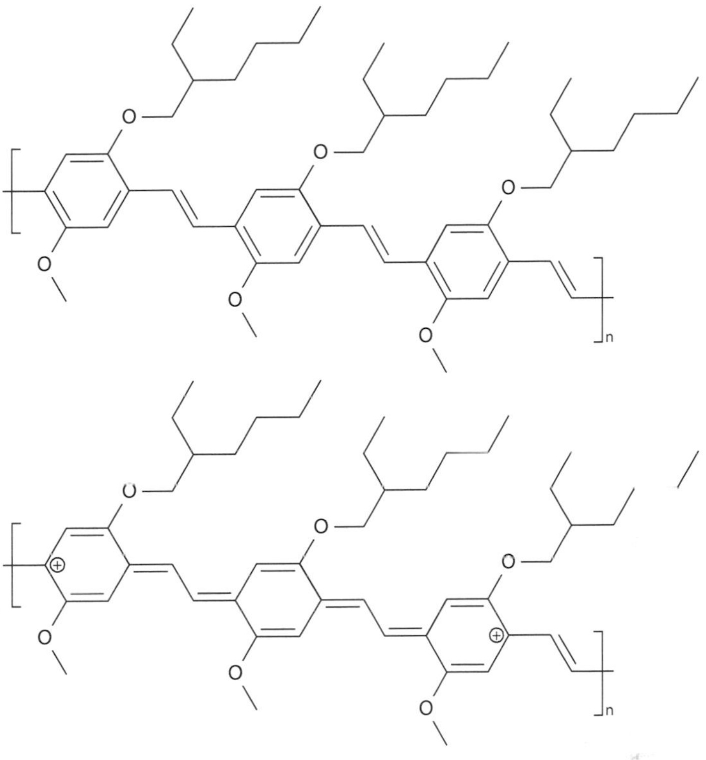

FIGURE 6.3 Molecular structure of MEH-PPV (above) and oxidized MEH-PPV (below) showing the change in structure that is induced through electrochemical doping.

involves multiple oxidation or reduction events at a single site on the polymer chain. The species created in these electrochemical reactions are highly reactive and sensitive to the presence of oxygen and water. For example, in poly[2-methoxy-5-(2'-ethyl-hexyloxy)-1,4-phenylene vinylene] (MEH-PPV), one of the most commonly used emissive polymers for LEC applications, it has been shown that over-oxidation occurs at only about 0.5 V above the voltage required to initially inject holes into the polymer.[17] The over-oxidation event in these materials results in an irreversible electrochemical reaction and a complete decomposition of the material. Similar results have been shown in a variety of other conjugated polymer structures of relevance to the field of iontronics as well. It is thought that controlling the counterion distribution in the device could prevent large concentrations from building up at the electrode, thereby discouraging overdoping reactions during device operation.[18] Fixed junction LECs should therefore allow for operation at higher applied bias for longer time periods, thereby increasing the operational lifetime and achievable brightness of traditional LECs. Indeed, fixed junction LECs demonstrate improved lifetimes over their dynamic counterparts in some cases.[7,10,15,21,22] However, in many cases, the overall performance of these fixed junction devices is qualitatively different from dynamic device behavior, making direct comparison of lifetimes under similar conditions (such as temperature, voltage, and brightness) difficult.

Perhaps the motivation for creating a fixed junction with the highest potential impact on the commercialization of LECs is the possibility that fixing the counterion distribution could allow for the operation of the LEC under reverse bias. As discussed previously, when the applied bias is reversed in a dynamic LEC, the counterions will move to opposite electrodes, thereby establishing a forward bias junction at opposite electrodes (oriented in the opposite direction). In this sense, a reversed biased junction is not a stable system in a dynamic junction LEC (see Figure 6.2). However, a junction that is stable under reverse bias would allow for a number of improvements over dynamic junction operation. First, the use of LECs in a practical display technology requires the ability to establish rectification (a diode-like current-voltage relationship) in order to avoid cross talk in an active matrix display architecture. For typical displays, a high rectification ration is necessary to be effective. Dynamic LECs sometimes do display a slight rectification, but typically, the current-voltage characteristics are symmetric (representing a rectification ratio of 1). In some fixed junctions, rectification ratios of up to 100 have been reported, bringing the performance much closer to that required for commercial display applications once lifetime and brightness in these systems are improved.[23,24]

In addition to potential improvements in the application potential for light-emitting structures, the establishment of a *pn* junction that is stable under reverse bias is necessary in order to achieve an appreciable photovoltaic response. A limitation common to traditional polymer photovoltaic devices based on a metal-insulator-metal device architecture is the short exciton diffusion length in these materials (5 to 20 nm). In general, a photogenerated exciton must diffuse to an interface (either donor/acceptor or polymer/electrode) in order to be separated and collected. If there is no interface, in general, within the diffusion length from where it was created, the exciton will recombine and be lost. In these devices, therefore, a tradeoff must be made between total light absorption and the efficiency of energy conversion—thicker films are capable of absorbing more light but result in a large number of excitons lost to recombination. This is most clearly observed in single layer polymer photovoltaic devices, where films must be very thin to exhibit a reasonable photovoltaic effect. Therefore, without a significant built-in potential, an LEC structure on its own will display little to no photovoltaic response. Employing a bulk heterojunction assists in improving efficiency by creating a larger interfacial area for exciton dissociation. However, the open-circuit voltages in these devices remain low, and the devices still exhibit a marked sensitivity to active layer thickness. The use of a *p-i-n* junction based device architecture, such as those demonstrated in the fixed-junction polymer LEC, could lead to a built-in potential within the polymer film, thereby potentially improving charge carrier collection in a photovoltaic device by serving to more effectively separate photogenerated charges in the bulk film. Using a fixed junction LEC device architecture, it may be possible to improve charge carrier collection efficiency by improving open circuit voltages without a compromise in overall light absorption. The improvement in photovoltaic response using a fixed, rather than dynamic junction LEC, and the improvement in the open circuit voltages of these devices over traditional polymer photovoltaic devices, has indeed been shown in several systems, although continued effort will be necessary to improve the overall performance of these devices for future applications.

Finally, a less applied motivation for creating a fixed junction LEC is that immobilization of the counterions in an LEC architecture can allow for the decoupling of the ionic and electronic contributions to various parameters of device operation, thereby providing a system that may allow future studies with the goal of achieving a deeper fundamental understanding of the operation of iontronic devices as a fundamental class of devices. For example, fixed junction devices may allow for characterization studies that are not compatible with simultaneous application of an external bias. One significant example of this approach is in the effort to elucidate the long debated operating mechanism of the LEC. Using the fixed junction approach, two groups have indeed been able to image electric field and potential profiles within a working device using scanning probe microscopic techniques, thereby giving direct evidence for the processes believed to be responsible for the observed behavior of LECs.[5,6,25]

Overall, the search for a robust method for fixing the junction in an LEC has taken two directions, either by controlling ionic motion through the temperature dependence of the ionic mobility in mixed conducting polymer films, or by taking advantage of covalent bonding, induced either synthetically or after the device had been assembled. Here we will discuss the methods that have been demonstrated to date.

6.3 DEMONSTRATED METHODS FOR CREATING FIXED JUNCTION LECS

The first demonstrated method for controlling the distribution of counterions in a polymer LEC took advantage of the temperature-dependant ionic conductivity in the polymer film. LECs are typically made using a single-layer polymer film that consists of the emissive polymer mixed with a solid-state ion-conducting polymer, usually poly(ethylene oxide) (PEO). At temperatures lower than the glass transition temperature of PEO, these blended polymer films show low ionic mobility. Gao et al. first showed that LECs could be cooled to approximately 100 K under an applied bias to achieve a fixed junction.[21] Devices operated at reduced temperatures demonstrated reduced turn-on time and a rectified current-voltage curve. Later, it was shown that such fixed junction behavior can be obtained up to just below the glass transition temperature at 200 K, although this appears to reduce the robustness of the fixed junction with respect to high applied bias.[11] As an extension to this work, Gao et al. showed that such devices could be operated as photovoltaic devices as well by blending two polymers in place of the single active species in order to achieve a blended bulk heterojunction device.[26] The devices prebiased in frozen junction mode showed an improvement in photovoltaic response, with significant values for open circuit voltage and moderate short circuit currents. One drawback of the initial frozen junction LECs, however, was the tendency to phase separate, resulting in the need to operate at temperatures lower than glass transition temperature (T_g) of PEO to ensure complete ion freeze out. Therefore, while this initial system showed promise in terms of fixing the junction successfully and brought attention to the potential advantages of fixed junction operation, clearly a system in which the normal operation of the device is possible at room temperature would be preferred.

An alternative, but similar approach was demonstrated by Yu et al. in which a solid-state ion conductor based on crown ether, with its glass transition temperature higher than room temperature, is used instead of PEO.[27] The device is first heated to above the glass transition temperature of the crown ether to around 60 to 80°C followed by cooling under an applied bias. The device can then be operated at room temperature as a fixed junction device. These systems performed similarly to the frozen junction devices, but remained fixed at room temperature.[12,28] In addition, these devices were able to demonstrate a similar photovoltaic response to the frozen junction devices at room temperature. Alternatives to the crown ether have also been demonstrated with melting points above room temperature, including several imidizole-based compounds (molten salts).[29] Unfortunately, at room temperature the junctions formed using this method did not remain completely fixed when a large enough applied bias was applied primarily due to resistive heating. A combined approach was demonstrated by Edman et al., in which the ionic conductivity in the ionic liquid devices are further reduced by operating the devices at reduced temperatures.[30] This work indeed showed that the junction created with the ionic liquid, while reversible under some conditions at room temperature, can be fixed at reduced temperatures. These devices demonstrate similar properties as seen in frozen junction devices, but with the potential improvement in emission and film morphology of devices made with ionic liquids in place of the standard blend with PEO.

An alternative approach to achieving a fixed junction LEC via the temperature dependence of ionic mobility was demonstrated by the construction of a self-assembled, chemically fixed homojunction in a semiconducting polymer device.[24] The architecture is essentially an LEC that contains polymerizable mobile counterions. Concurrently with the electrochemical doping of the conjugated polymer, the electrochemical generation of radical anions and cations in the conjugated polymer initiates covalent bonding via the vinylic groups of the ionic monomers. This process immobilizes the counterions, preventing reversibility of the electrochemical doping and stabilizing the established homojunction. The resulting devices demonstrate high rectification, unipolar light emission, a linear relationship between current and radiance, and stability at room temperature under a wide range of operating voltages. This approach was also applied to photovoltaic devices in a single-layer, single active component as well as blended bulk heterojunction devices.[31] This method showed strong potential for improving the state-of-the art in organic photovoltaic systems due to high open circuit voltages; however, the overall performance of the system is still low in comparison with traditional polymer photovoltaic devices. Improvement in this system is expected with continued development of ion-paired monomers with improved film forming properties.

Another class of devices that are capable of fixed junction operation are those based on transition metal complexes such as $[Ru(bpy)_3]^{2+}$.[32] These materials are ionic in nature and when put into thin film form have only one mobile counterion. Although the cationic species remains fixed during application of an external bias, because the anion remains mobile the device remains essentially dynamic and the device characteristics essentially symmetric. These materials have been pursued in depth in LEC structures since 1996 due to their bipolar electrical characteristics and high luminescence efficiency. The motivation for creating fixed junctions in these materials has been motivated primarily by the effort to overcome the characteristically slow turn-on times in these materials that result from low intrinsic ionic mobility. The first demonstration of devices in which the anion is immobilized during operation was published in 1996.[33] High ionic mobility in this case was achieved via swelling of the thin film during charging. Dry films then demonstrated drastically improved turn-on times and no loss of overall performance. However, these devices tended to remain fixed only within a small range of operating conditions. For example, elevated operation temperatures or sustained reverse bias could result in reversal of the fixed counterion distribution. An alternative approach reported in 1997 took advantage of the temperature-dependent ionic mobility in these films in much the same way as reported in so called "frozen-junction" polymer LECs.[34] These devices similarly showed promising characteristics but an inherent instability. A unique approach to fixing the anion distribution in transition metal LECs has also been reported more recently. In 1998, Elliott et al. demonstrated devices in which the active material is a polymerizable ruthenium complex.[35] In this study, the materials were polymerized during device operation. The polymerization of the active materials tended to reduce the ionic mobility of the film and resulted in a fixed ion distribution under normal operating conditions to the extent explored.

Conjugated polymers with built-in ionic pendant groups have also been used in LEC structures using a similar approach as seen in the transition metal devices. In these materials, similar approaches to fixing the distribution of counterions can also be envisioned. However, one unique approach reported in 2003 takes advantage of the unusual temperature-dependent properties of the emissive material itself.[22] In the PFN+Br- material, an amorphous phase is seen at room temperature. However, at elevated temperatures the material tends to crystallize, thereby reducing ionic mobility in the films. Therefore, devices can be charged at room temperature, where the amorphous phase allows easy establishment of the desired counterion distribution, and fixed by increasing the temperature of the system and crystallizing the active material. These devices showed improved turn-on behavior and lifetime.

An innovative technique reported by the Lonergan group extends the use of the conjugated ionomers to create a junction that is not only fixed through simple immobilization of counterions, but which removes the counterions from multilayered structures to create a device that resembles an inorganic *pn* junction.[23] A multilayered structure is utilized here in which the active material in each layer consists of an anionic and cationic pendant group, respectively. The counterions to these groups are washed away via solvent soaking, leaving a system in which the charged layers are truly immobile. These devices were shown to have highly rectified current voltage characteristics and stability over long operation periods. Although the materials used here were not practical for many applications, the approach used is promising for a range of systems. Using a similar approach, Bernards et al. demonstrated a two-layer system consisting of DPAS- Na+ and $(Ru(bpy))_3^{2+}(PF_6^-)^{2.6}$. In this system, the mobile sodium and hexafluorophosphate ions are retained while the conjugated structures remain fixed. It is this fixed junction between the immobile DPAS- and $Ru(bpy)_3^{2+}$ that leads to rectification and photovoltaic behavior. Bernards and coworkers were also able to use this system to successfully elucidate the electric field distribution within the device using scanning probe techniques as discussed previously.

Another method for carefully controlling the counterion distribution and fixing it using chemical techniques was presented in 2007 by Simon et al.[37] In this system, charged self-assembled monolayers were used to create a fixed distribution of charges in a conjugated polymer film. In this system, significant improvements were observed both in the efficiency of light emission as well as photovoltaic behavior. This approach, together with the conjugated polymer *pn* junctions reported by Cheng et al.[23] and the chemically fixed junction technique reported at about the same time, form the primary focus of research into fixed-junction LEC approaches being pursued in the last decade. These chemical techniques not only provide a way to control the location of counterions in a more stable way, but also in these systems, there exists the possibility to control the doping distribution within the film and other aspects of the electrochemistry as well.

6.4 CONCLUSIONS

Creating a fixed electronic junction in an organic device has important potential advantages over metal-insulator-metal devices including improved performance and application potential. A range of creative approaches to this problem has been proposed and demonstrated with varying levels of success. Implementation of these approaches on a commercial level will require significant improvements in performance on several metrics, however. Such improvements will likely come from advances in our fundamental understanding of ionic carriers and electrochemistry of conjugated organic structures. It is likely that with such improvements, multijunction systems will be realized on a major scale in future technologies.

REFERENCES

1. Q. Pei, G. Yu, C. Zhang, Y. Yang, A.J. Heeger, *Science* 1995, *269*, 1086.
2. Q. Pei, Y. Yang, *Synth. Met.* 1996, *80*, 131.
3. J.C. DeMello, N. Tessler, S.C. Graham, R.H. Friend, *Phys. Rev. B* 1998, *57,* 12951.
4. J.M. Leger, B. Ruhstaller, S.A. Carter, *J. Appl. Phys.* 2005, *98*, 124907.
5. L.S.C. Pingree, D.B. Rodovsky, D.C. Coffey, G.P. Bartholomew, D.S. Ginger, *J. Am. Chem. Soc.* 2007, *129*, 15903.
6. J.D. Slinker, J.A. DeFranco, M.J. Jaquith, W. Silveira, Y.W. Zhong, J.M. Moran-Mirabal, H.G. Craighead, H.D. Abruña, J.A. Marohn, G.G. Malliaras, *Nature Mater.* 2007, *6*, 894.
7. L. Edman, *Electrochim. Act.* 2005, *50,* 3878.
8. Y. Yang, Q. Pei, *Appl. Phys. Lett.* 1996, *68*, 2708.
9. F.P. Wenzyl, L. Holzer, S. Tasch, U. Scherf, K. Mullen, B. Winkler, A.W.H. Mau, L. Dai, G. Leising, *Synth. Met.,* 1999, *102*, 1138.
10. L. Edman, D. Moses, A.J. Heeger, *Synth. Met.,* 2003, *138*, 441.
11. J. Gao, Y. Li, G. Yu, A.J. Heeger, *J. Appl. Phys.* 1999, *86*, 4594.
12. L. Edman, M. Pauchard, D. Moses, A.J. Heeger, *J. Appl. Phys.* 2004, *95*, 4357.
13. F. Habrard, T. Ouisse, O. Stephan, M. Armand, M. Stark, S. Huant, E. Dubard, J. Chevrier, *J. Appl. Phys.* 2004, *96*, 7219.
14. J.H. Shin, S. Xiao, L. Edman, *Adv. Funct. Mater.* 2006, *16*, 949.
15. Y. Shao, G.C. Bazan, A.J. Heeger, *Adv. Mater.* 2007, *19*, 365.
16. Y. Kervella, M. Armand, O. Stephan, *J. Echem. Soc.* 2001, *148*, 155.
17. A.L. Holt, J.M. Leger, S.A. Carter, *J. Chem. Phys.* 2005, *123*, 44704.

18. J. Fang, P. Matyba, N.D. Robinson, L. Edman, *J. Am. Chem. Soc.* 2008, *130*, 4562.
19. J.L. Bredas, G.B. Street, *Acc. Chem. Res.* 1985, *18*, 305.
20. D. Baeriswyl, D. Campbell, S. Mazumdar, *Conjugated Conducting Polymers,* H. Keiss, Ed., Springer, Berlin, 1992.
21. J. Gao, G. Yu, A.J. Heeger, *Appl. Phys. Lett.* 1997, *71*, 1293.
22. L. Edman, M. Pauchard, B. Liu, G. Bazan, D. Moses, A.J. Heeger, *Appl. Phys. Lett.* 2003, *82*, 3961.
23. C.H.W. Cheng, M.C. Lonergan, *J. Am. Chem. Soc.* 2004, *126*, 10536.
24. J.M. Leger, D.B. Rodovsky, G.P. Bartholomew, *Adv. Mat.* 2006, *18*, 3130.
25. J.C. DeMello, J.J.M. Halls, S.C. Graham, N. Tessler, R.H. Friend, *Phys. Rev. Lett.* 2000, *85*, 421.
26. J. Gao, G. Yu, A.J. Heeger, *Adv. Mat.* 1998, *10*, 692.
27. G. Yu, Y. Cao, M. Andersson, J. Gao, A.J. Heeger, *Adv. Mat.* 1998, *10*, 385.
28. L. Edman, M.A. Summers, S.K. Buratti, A.J. Heeger, *Phys. Rev. B* 2004, *70*, 115212.
29. C. Yang, Q. Sun, J. Qiao, Y. Li, *J. Phys. Chem. B* 2003, *107*, 12981.
30. J.H. Shin, S. Xiao, A. Fransson, L. Edman, *Appl. Phys. Lett.* 2005, *87*, 43506.
31. J.M. Leger, D.G. Patel, D.B. Rodovsky, G.P. Bartholomew, *Adv. Funct. Mater.* 2008, *18*, 1212.
32. J. Slinker, D. Bernards, P.L. Houston, H.D. Abruna, S. Bernhard, G.G. Malliaras, *Chem. Comm.* 2003, 2392.
33. K.M. Maness, R.H. Terrill, T.J. Meyer, R.W. Murray, R.M. Wightman, *J. Am. Chem. Soc.* 1996, *118*, 10609.
34. K.M. Maness, H. Matsui, R.M. Wightman, R.W. Muray, *J. Am. Chem. Soc.* 1997, *119*, 3987.
35. C.M. Elliott, F. Pichot, C.J. Bloom, L.S. Rider, *J. Am. Chem. Soc.* 1998, *120*, 6781.
36. D.A. Bernards, S. Flores-Torres, H.D. Abruña, G.G. Malliaras, *Science* 2006, *313*, 1416.
37. D.T. Simon, D.B. Stanislowski, S.A. Carter, *Appl. Phys. Lett.* 2007, *90*, 103508.

7 Electrochromic Displays

Magnus Berggren, David Nilsson,
Peter Andersson Ersman, Payman Tehrani,
and Lars-Olov Hennerdal

CONTENTS

7.1 INTRODUCTION

Information, such as graphics and text, is preferably displayed along the surface of a large, thin, and flexible carrier. Major efforts were devoted during the 19th and 20th centuries to develop reel-to-reel manufacturing and printing techniques[1] to produce magazines, billboards, etc. on paper. The goal has always been to achieve a technology that enables rapid distribution of information in our society. In the second revolution of information technology, digital signal processors, electronic displays, and mobile communication systems have been developed that together have considerably shortened the delay time from text and graphics production to the actual reading events. There is a great interest to combine the form factor of the easily handled paper substrates and the features of advanced electronic communication and displaying systems. Thus, it has proven very challenging to merge the two technologies, which in part is due to incompatibility regarding materials and the philosophy of manufacturing techniques.

During the last two decades, organic electronic materials have been explored as the active materials in a vast array of different electronic devices, such as for displays, transistor circuits, and solar cells. Organic materials can be processed and patterned from solutions to form device systems on a wide range of flexible surfaces. This makes it possible to print entire electronic systems on, for instance, plastic foils and paper.[2] In addition, it would be of great interest to produce electronics on paper carriers in order to make electronics generally easier to recycle. Electrochromic (EC) displays[3] based on organic materials are a class of devices that exhibit suitable properties and features that promise for a future electronic paper or paper electronics technology.

EC denotes the phenomenon when the optical absorption characteristics, that is, the color, of a material are controlled when a burst of charge is applied. Often, such active control originates from switching the electrochemical state of the material.

In this chapter, some of the characteristics of EC of polythiophenes are reviewed. Particular focus is given to device architectures, optical and electronic material properties, and applications of EC in paper display applications.

7.2   ELECTROCHROMIC DEVICE CONFIGURATIONS

In its simplest form, an EC display cell is comprised of the actual display electrode including the EC material, a counter electrode, and a common electrolyte, together with necessary electrical wiring and a power source. Some conjugated polymers exhibit high enough electronic conductivity in at least two oxidation states, which makes supporting electrodes and conductors unnecessary. In other words, the conjugated polymer film can both define the EC display functionality and serve as its own electrode material. Poly(3,4-ethylenedioxythiophene) (PEDOT)[4] doped with poly(styrenesulfonate) (PSS) is one such kind of material and has been exploited in EC devices since the early 1990s.[3] PEDOT:PSS exhibits many desirable properties that are suitable for large area and flexible electronics and the material has therefore been utilized in several single-use electronics applications.[5]

PEDOT:PSS can be coated onto foils and fine paper using standard techniques originally developed for the foil and paper industry, see Figure 7.1. Such homogenous conducting coatings exist today and are provided by industry[6] (see Figure 7.1). PEDOT:PSS electrodes can be patterned utilizing subtractive printing techniques, such as depositing lines of oxidants or using electrochemical over-oxidation deactivation.[7] Electrolytes can be deposited using, for instance, inkjet or screen printing.[8] Solidified electrolytes can be defined using, for instance, polyethylene glycol as the gelling agent including an aqueous salt electrolyte. For EC displays based on PEDOT:PSS, electrochemical switching is governed by the following reversible electrochemical reaction

$$PEDOT + PSS:M \rightarrow PEDOT^{n+}:PSS^{n-} + nM^+ + ne^- \qquad (7.1)$$

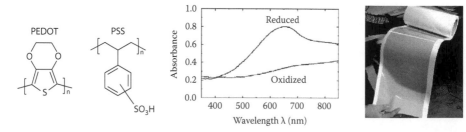

FIGURE 7.1   The molecular structure of PEDOT and PSS:H (left), the optical absorption spectrum of a thin film of PEDOT:PSS (right) in which PEDOT has been switched to its neutral and oxidized state (middle), respectively. PEDOT:PSS coatings deposited on polyethylene-coated fine paper.

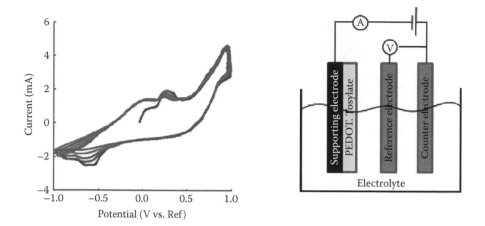

FIGURE 7.2 The cyclic voltammetry spectrum for PEDOT (left) and the three-electrode electrochemical setup (right).

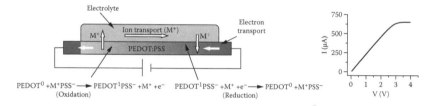

FIGURE 7.3 The S1 configuration. An electrochemical gradient is formed along the PEDOT:PSS electrode, which gives that the PEDOT phase closest to the negatively addressed side becomes relatively more neutral (low conductivity). As the voltage is increased, this part of the PEDOT:PSS electrode turns more and more neutral, which gives that the steady-state electronic current pinches off (right). Instead, PEDOT closest to the positively addressed side turn more oxidized.

that involves electrons (e^-) and protons or metal ions (M^+). In a three-electrode setup, the oxidation potential can be predicted from the cyclic voltammetry (CV) curve (see Figure 7.2).

The oxidation potential of PEDOT[9] is found to be around 0.2 vs. the AG/AgCl reference electrode (see Figure 7.2).

If one continuous electrode film is in part coated with an electrolyte layer, the so-called *Structure 1* (S1) configuration is formed (see Figure 7.3).[10] As both sides are electrically biased of this PEDOT:PSS electrode film a current flows along it, which is associated with that a potential gradient is established. This potential gradient results in an electrochemical gradient along the entire PEDOT:PSS film that is in contact with the electrolyte. The PEDOT phase then continuously changes its oxidation state from relatively more neutral to relatively more oxidized (see Figure 7.3). As the voltage difference that is applied to the two sides increases, the scalar value of the gradient increases. First at low voltages, the electronic current increases linearly

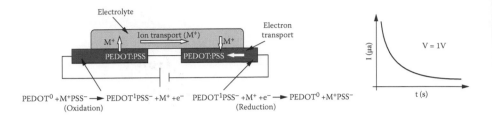

FIGURE 7.4   The S2 configuration (left) and its current vs. time characteristics (right).

vs. the voltage. At higher voltages, typically above 1 V, the current evolution is suppressed by a typical pinch-off behavior. This is because the reduction toward a more neutral state of PEDOT gives an overall increase of the impedance along the electrode (see Figure 7.3).

A PEDOT:PSS electrode can be divided into two separate adjacent electrodes (*Structure 2, S2*), which have a very different kind of device behavior (see Figure 7.4), as compared to S1. In the S2 configuration, the electronic current is interrupted and ideally, every charge that is injected to or withdrawn from the two electrodes is counterbalanced by an electrochemical reaction occurring at the two electrodes.

The current vs. time characteristics for a specific bias voltage exhibit typically a monotonically decreasing behavior (see Figure 7.4). As the current approaches zero or the level of the inherent leakage current, the PEDOT phase of the positively addressed electrode becomes eventually homogenously fully oxidized ($PEDOT^{n+}$) and PEDOT in the negatively addressed electrode becomes reduced to the neutral state ($PEDOT^0$).

Depending on the relative sizes of the two electrodes, the relative capacity for reduction vs. oxidation can be controlled. For PEDOT:PSS, the level of oxidation is less than 30%. So for a PEDOT:PSS two-electrode electrochemical device cell, complete reduction of one of the electrodes requires that the counter electrode be of considerably larger volume, typically a factor of 5 or so.

7.3   ELECTROCHROMIC COLORS AND CONTRASTS

In an EC display, based on the S2 configuration, the maximum color change that can be achieved is predicted by the absorption characteristics of the EC material switched to its different oxidation states together with the current switching capacity of the device cell. The latter implies that the counter electrode and the display electrode must be balanced in terms of volume (accessible sites for oxidation and reduction) and capacitance so that the display electrode can be switched to its fully oxidized, reduced, or neutral states. For PEDOT:PSS-based display cells (see Figure 7.5), the color switch contrast is given by the color of PEDOT in its neutral and oxidized state, respectively. In the neutral state, PEDOT absorbs mainly at red to yellow wavelengths giving that the material appears as a strong blue colorant. In the oxidized state, PEDOT turns almost transparent with a weak sky-bluish hue. PSS appears completely transparent at visible wavelengths. In this

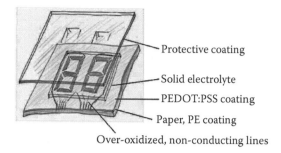

Protective coating

Solid electrolyte

PEDOT:PSS coating

Paper, PE coating

Over-oxidized, non-conducting lines

FIGURE 7.5 A seven-segment EC display manufactured on polyethylene-coated fine paper (left) and its device architecture (right).

S2 display configuration, the PEDOT:PSS display electrode serve as a dynamic and translucent (non-opaque) colorant absorbing incoming and scattered light along the white paper surface. In other words, PEDOT:PSS EC electrodes function as a dynamic absorption shutter to modulate scattering of white light along the paper coating.

In many display applications, the contrast is of utter importance. By reaching a high contrast, the overall appearance of the information and graphics content that is displayed is improved and appears as clear. For reflective displays, the contrast can be defined as

$$\Delta E^* = \sqrt{\left(\Delta L^*\right)^2 + \left(\Delta a^*\right)^2 + \left(\Delta b^*\right)^2} \tag{7.2}$$

where L^*, a^*, and b^* are the color coordinates and ΔE^* is the optical switch contrast. For a PEDOT:PSS-only paper display, the switch contrast ΔE^* equals a value of 29 for an optimal film thickness of the active PEDOT:PSS layer (t = 400 nm). To improve the EC switch contrast, several options are available such as blending in different other EC materials into the PEDOT:PSS system or just stacking additional EC layers or devices on top of each other. In any case, it is important that the combined materials truly are complementary to each other from a spectral point of view and that they all can appear transparent at the same time and bias condition in the device.

PProDOT-Hx$_2$[11] is an EC material that exhibits strong optical absorption at blue-green-yellow wavelengths in its neutral state. In its oxidized state, the material appears almost completely transparent (see Figure 7.6). Together, these features make PProDOT-Hx$_2$ an excellent complementary EC colorant to PEDOT in order to achieve high optical switching contrasts. By adding only an 80-nm thin film of PProDOT-Hx$_2$ on top of the PEDOT:PSS display electrode before the electrolyte is deposited, the optical switch contrast (ΔE^*) increases from 29 to 54 (see Figure 7.6).

Contrast is not the only important device feature of EC reflective displays; one needs pure colors as well. If one compares different polythiophene derivatives, one finds that it is easy to modify the optical absorption spectrum, that is, the color, of these materials in their neutral state. This can be achieved simply by including

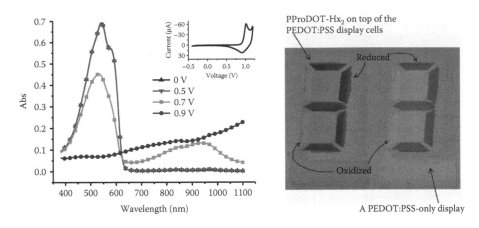

FIGURE 7.6 The optical absorption characteristics of a thin film of PProDOT-Hx$_2$ (left), coated onto an ITO electrode, at different voltages vs. the Ag/AgCl reference electrode. The resulting improvement achieved by including PProDOT-Hx$_2$ on top of the PEDOT:PSS electrode in a seven-segment paper display (right).

different kinds of substituents and molecular groups along and inside the polythiophene main chain. Two examples are given in Figure 7.7. However, to achieve a dynamic colorant that can be switched reversibly, while included as the active material in an electrochemical cell in-between desired translucent and colored states, one also needs proper conductivity of the involved charged species. This suggests that fast electrochemical switching of an entire EC polymer bulk requires fast migration of both the electronic and ionic charge carriers at all oxidation states. For ionic migration, fast conduction can be obtained by adding ethylene oxide (CH_2-O-CH_2) groups as the substituents along the conjugated polymer chain. For electronic charge migration, high electronic conductivity of the polymer in both the neutral and oxidized state can be reached by including EDOT-groups inside the polythiophene main chain (see Figure 7.7). WM 144 includes both ethylene oxide side groups and EDOT main chain groups and exhibits fast EC switching and was successfully used as an EC material in EC displays.

7.4 DISPLAY SYSTEMS—ELECTROCHROMIC TFT DISPLAYS

As PEDOT is switched in-between the oxidized and neutral state, a conductor-semiconductor transition occurs. Thus far, only the associated color-switching feature has been discussed in this chapter. The switching is also associated with a tremendous change of the electronic hole conductivity within the PEDOT phase of the material system. This results in electrochemical control of the impedance along the PEDOT:PSS film and can be utilized to make electrochemical transistors.[10,12] These electrochemical transistors can be built up using the very same active materials as those that are included in the EC display cells. This offers us the opportunity to realize transistor-display electronic systems, that is, smart pixels made from a few different materials, to achieve large area and pixilated display systems on, for

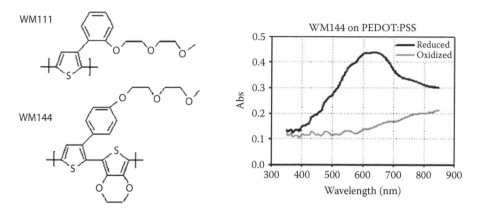

FIGURE 7.7 Two polythiophene derivatives (left) that only differ by the EDOT-group included in the main chain. Both polymers provide good ionic conductivity but only the WM 144 offers good enough electronic conductivity in both the neutral and oxidized states. The optical absorption spectrum of WM 144 spin coated onto an ITO electrode (right). Despite the many similarities between the two materials, WM 111 has never worked properly in EC displays due to poor electronic conductivity.

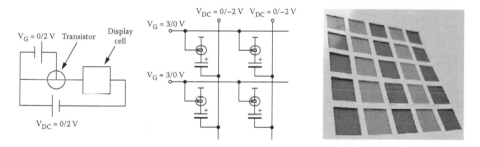

FIGURE 7.8 The principal circuit of an electrochemical smart pixel (left) and the layout of an EC matrix display (center). An image of a 5 × 5 matrix EC display (right) in which PEDOT of the display electrodes in the first, third, and fifth columns are all switched to the (right) reduced state. The PEDOT phase of the other display electrodes are all in their semi-oxidized (transparent) state.

instance, paper.[13] In a thin-film transistor (TFT)-driven LCD display, the transistors are utilized for addressing and updating each individual display cell. In each wire crossing, a so-called smart pixel is included that combines one display cell with one or several addressing transistors. The transistor provides the updating signal and information non-volatility of each display cell. As the transistor is open, that is, the channel of the transistor is in its high conducting state, the coloring and de-coloring of the display cell is fast. As the transistor is closed, the display cell is decoupled from the addressing columns and rows, which gives that the information content of the display cells is retained (see Figure 8.8). In the electrochemical display systems, each electrochemical transistor along one row is opened to enable updating of all the display cells along this particular row. The updating signal current is applied along the column wires.

7.5   MANUFACTURING OF EC DISPLAYS

EC displays are certainly a good choice for future paper electronics technology. In order to fully merge the commonly used paper, being an easily distributed and cellulose-based information carrier, with electronics it is crucial that all included devices can be manufactured using standard conversion and printing technology. Specifically, to manufacture EC displays based on PEDOT:PSS, one needs to deposit and build up a device architecture combining the EC PEDOT:PSS material, a solid electrolyte layer providing fast electrochemical switching and a protective sealing layer. Homogenous coatings of PEDOT:PSS on plastic foils exists today in the form as the Orgacon foil, also coated on paper. This technology can easily be applied also to PE-coated paper (see Figure 7.1). The PEDOT:PSS coated paper then needs to be subtractively patterned to form non-conducting lines that demarcate conductors, EC display cells, and counter electrodes.

Local de-activation of PEDOT:PSS can be realized using the so-called electrochemical over-oxidation patterning process.[7] In this process, an aqueous electrolyte is locally deposited onto the PEDOT:PSS coating. Then, an external electrode is applied to the electrolyte patterns. As this electrode is grounded and the PEDOT:PSS coating is exposed to an anodic over-potential (typically above 5 V), the PEDOT phase that is in direct contact with the electrolyte becomes electrochemically over-oxidized. Screen-printing has proven to be a successful technique to achieve such electrochemical over-oxidation pattering of large area PEDOT:PSS coatings, for instance coated on paper. In this case, the electrolyte is printed onto the PEDOT:PSS layer through the screen printing mesh and a metal squeegee serves as the grounded electrode. A conducting reel is then utilized as the second electrode in order to apply an anodic signal to the PEDOT:PSS coating. For a seven-segment display, the counter electrodes are commonly placed along the two "dead" inner areas surrounded by all the actual display segments (see Figure 7.5).

In the configuration presented here, the counter and display electrodes are placed next to each other in the same plane. This gives that the ion migration takes place primarily along the plane parallel to the carrying surface. This forces us to deposit a thick electrolyte layer in order to achieve fast switching of the display cells. Typically, the gelled and solid electrolyte needs to be at least 30 μm thick. From a printing point of view, only screen-printing can be used to deposit such thick layers of electrolytes. As sealing layers, different kinds of lacquers and self-adhesive foils have been explored and those are favorably deposited or glued onto the top electrolyte immediately after the screen-printing steps.

At the Printed Electronics Arena laboratory (PEA), Acreo and Linköping University together with partner industry jointly developed an all-in-line printing process for EC paper displays.[14] A label printer (Nilpeter FA3300/5) has been modified to include a lineup of the patterning steps described previously. Since cylinder-based and all-in-line machines are chosen, flat screen-printing processing cannot be utilized. This particular label printer includes an array of rotary screen printing steps combined with flexo- and coating stations (see Figure 7.9).

FIGURE 7.9 The label printer, at PEA, used to manufacture EC paper displays (left). In addition, an integrated electrochemical paper label, including push buttons, EC displays, printed batteries, and electrochemical transistors (right).

REFERENCES

1. Kipphan, H., *Handbook of Print Media Technologies and Production Methods*. Springer, New York, 2001.
2. Eder, F. et al., Organic electronics on paper. *Applied Physics Letters* **84**(14), 2673 (2004); Andersson, P. et al., Active matrix displays based on all-organic electrochemical smart pixels printed on paper. *Advanced Materials* **14**(20), 1460 (2002).
3. Pei, Q., Zuccarello, G., Ahlskog, M., and Inganäs, O., Electrochromic and highly stable poly(3,4-ethylenedioxythiophene) switches between opaque blue-black and transparent sky blue. *Polymer* **35**(7), 1347 (1994).
4. Groenendaal, L. et al., Poly(3,4-ethylenedioxythiophene) and its derivatives: past, present, and future. *Advanced Materials* **12**(7), 481 (2000); Heywang, G. and Jonas, F., Poly(alkylenedioxythiophene)s—New, very stable conducting polymers. *Advanced Materials* **4**(2), 116 (1992).
5. Zhu, Z.-T. et al., A simple poly(3,4-ethylene dioxythiophene)/polystyrene sulfonic acid) transistor for glucose sensing at neutral pH. *Chemical Communications* 1556 (2004).
6. Louwet, F., Los Angeles, CA, 2008 (unpublished).
7. Tehrani, P. et al., Patterning polythiophene films using electrochemical over-oxidation. *Smart Materials and Structures* **14**(4), 21 (2005).
8. Jeong Ho, C. et al., Printable ion-gel gate dielectrics for low-voltage polymer thin-film transistors on plastic. *Nature Materials* **7**(11), 900 (2008); Tehrani, P. et al., Improving the contrast of all-printed electrochromic polymer on paper displays. *Journal of Materials Chemistry* **19**(13), 1799 (2009).
9. Otero, T.F. and Romero, M.C., UK, 2008 (unpublished).
10. Nilsson, D. et al., Bi-stable and dynamic current modulation in electrochemical organic transistors. *Advanced Materials* **14**(1), 51 (2002).
11. Cirpan, A. et al., Electrochromic devices based on soluble and processable dioxythiophene polymers. *Journal of Materials Chemistry* **13**(10), 2422 (2003).
12. Bernards, D.A and Malliaras, G.G., Steady-state and transient behavior of organic electrochemical transistors. *Advanced Functional Materials* **17**(17), 3538 (2007).
13. Andersson, P., Forchheimer, R., Tehrani, P., and Berggren, M., Printable all-organic electrochromic active-matrix displays. *Advanced Functional Materials* **17**(16), 3074 (2007).
14. Berggren, M., Nilsson, D., and Robinson, N.D., Organic materials for printed electronics. *Nature Materials* **6**(1), 3 (2007).

8 Conjugated Polymers as Actuators for Medical Devices and Microsystems

Edwin W. H. Jager

CONTENTS

8.1 INTRODUCTION

In conducting polymers, ion transport occurs during the (electro-)chemical oxidation and reduction of the polymer. This redox change results in a change of material properties such as conductivity [1], electrochromism [2], and wettability [3, 4]. The ion transport during this redox switching has also been used in drug release applications [5–7]. This reversible intercalation motion of the ions also results in a volume change of conducting polymers. These materials were proposed as actuator materials by Baughman et al. for their large strains and low operating potentials [8]. The first such conducting polymer-based actuator was demonstrated in the early 1990s by Pei and Inganäs [9]. The phenomenon was thereafter demonstrated by other laboratories [10–14] and the new field of conjugated polymer actuators emerged.* Initial research was focused on understanding the physical principles behind the electrochemically induced actuation and exploring the limitations.

8.2 ACTUATION IN CONJUGATED POLYMERS

As said, the electrochemical switching of conjugated polymers may result in a volume change of the material due to the insertion and extraction of counter ions into the polymer matrix. Depending on the dopant used in the polymer, two different redox reactions (and accompanying ion flows) are possible [11, 15–17]. In a general sense, the ion flow and redox reactions can be described as follows. For a polymer P doped with small, mobile anions (a⁻) in contact with an electrolyte containing both mobile cations and anions, the reaction is:

$$P^+(a^-) + e^- \leftrightarrow P^0 + a^-(aq) \tag{8.1}$$

That is, when reducing the polymer to its neutral state, anions a⁻ are expelled and when oxidizing the polymer, anions are inserted into the polymer matrix in order to compensate for the charge imbalance. On the other hand, for a polymer P doped with large, immobile anions A⁻ in contact with an electrolyte containing small mobile cations M⁺, the reaction is:

$$P^+(A^-) + M^+(aq) + e^- \leftrightarrow P^0(A^-M^+) \tag{8.2}$$

That is, cations M⁺ are inserted when the polymer is reduced and expelled when the polymer is oxidized. In the former case, the volume typically expands in the *oxidized* state, that is, when a positive potential is applied, and in the latter case the volume of the polymer expands in the *reduced* state, that is, when a negative potential is applied. In the former case, however, there may be two moving species because not only reaction (8.1) occurs, but also reaction (8.2) may occur, which can lead to a "twitching" behavior [16]. Thus, it is preferable to have only one moving species. Therefore, polypyrrole (PPy) doped with large immobile anions, such as dodecylbenzene sulfonate (DBS), has been employed frequently because it provides a smooth motion (with only cations as the moving species), stability, and long lifetime. In addition to the ion motion, osmotic flow of solvents due to the altered ion concentration inside the polymer matrix [18] and conformational changes and coulombic repulsion of the polymer chains may also contribute to the volume change [19, 20].

Initially, the volume change was estimated to be only a few percent [17]. The numbers were deduced by measuring the bending angle of a tri-layer device consisting of a polyethylene film, an evaporated Au layer, and an electrosynthesized PPy

---

* Conjugated polymers are also commonly referred to as electroactive polymers (EAP) when used as actuators. EAPs are commonly classified as electric or ionic, with conducting polymers being a subgroup within the latter. Other groups of ionic actuators are ion polymer metal composites (IPMCs), carbon nanotubes, and gels, but these fall beyond the scope of this chapter.

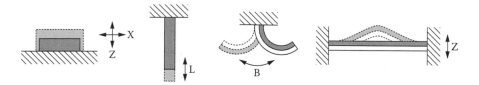

FIGURE 8.1   Different actuation modes used for conjugated polymer actuators, bulk expansion, linear strain, bending beam, and buckling sheet.

layer and using Timoshenko's bending beam theory to calculate the volume change. However, it was shown that the volume change is highly anisotropic. In situ volume change measurements using atomic force microscopy showed that the reversible volume change in the perpendicular direction could be up to 30 to 40% for 0.8 to 1.5 µm films of PPy(DBS) [21, 22]. These large volume changes were later confirmed for "thick" PPy (40 to 55 µm thick), showing expansions ranging from 4 to 19 µm [23]. While the volume change is not bistable, intermediate states can be reached with the amount of bending depending on the charge added to or removed from the polymer [24–26].

Conjugated polymers exhibiting this volume changing capacity can be used as an active material in actuators. This actuation can take a variety of forms (Figure 8.1). The bulk expansion can be used as a piston-like actuator; the linear strain for linear actuators such as stripes and tubes; and the bending motion as a rolling sheet or hinge in bi- or multilayer devices. A variant of the bending multilayer is a buckling configuration where at least two sides of the actuator sheet are clamped or attached.

8.3 MATERIALS AND FABRICATION

The material properties are determined by the monomer of which the conjugated polymer is constituted, the dopant ions trapped in the polymer, and synthesis conditions. Although most conjugated polymers are electrochemically active, only a few have been used as actuator materials (Figure 8.2). Predominantly, PPy is used [9–11,

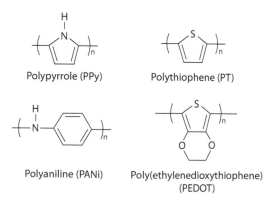

FIGURE 8.2   The chemical structures of some common conjugated polymers used as actuator materials.

13] because it is easy to fabricate and actuate. However, polyaniline [14, 27, 28] and various polythiophenes [9, 29] have also been used in actuators. Bohn et al. have demonstrated that a substituted form of PPy, that is, PEDOP, can actuate, although the performance was far less than that of ordinary PPy [30].

A film of conjugated polymers can be applied using several methods such as spin coating from a polymer solution, chemical polymerization, and electrochemical polymerization. For actuator applications, electrochemical polymerization is most commonly utilized. Electrochemical polymerization, a technique similar to electroplating, is a stable and well-reproducible synthetic method and one can rapidly synthesize materials with different properties by minor adjustment of the procedure. Typically, the polymer is synthesized from a solution containing the monomer and a salt (which becomes the dopant). Since the procedure relies on conduction/transfer of electrons, the conjugated polymer must be synthesized on a conducting surface, known as the working electrode (WE). This conducting surface can be a part of the actuator [9] or be used only during the synthesis. Otero et al. synthesized PPy on a stainless steel electrode, after which the PPy film was peeled off and laminated onto Scotch tape to form a bilayer or triple layer actuator [31]. The tube actuator as developed by the Wollongong group is synthesized on a solid Pt tube [32]. After PPy electrosynthesis, the Pt tube is removed and electrical contacts are added.

To complete the electrochemical circuit of the synthetic "cell", a counter or auxiliary electrode (CE) and (most often) a reference electrode (RE) are used, all connected to a power supply, such as a potentiostat or galvanostat. In general, a constant current (galvanostatic method) or constant potential (potentiostatic method) is applied to synthesize the material, although more elaborate deposition methods have been reported, such as a sequence of potential steps [25]. Table 8.1 lists some common recipes used.

TABLE 8.1
Common PPy Electrosynthesis Procedures

WE	Electrolyte	Method	Time	Temperature	Ref.
Pt	0.06 M pyrrole + 0.05 M TBA PF_6 in propylene carbonate	Galvanostatic 0.15 mA/cm^2	6 h	−28°C	[32]
Pt	0.2 M pyrrole + 0.1 M $LiClO_4$ in acetonitrile + 2% water	Consecutive square waves of potential −500 mV (2 s) and 800 mV (10 s) vs. Ag/AgCl		RT	[19]
Au	0.1 M pyrrole + 0.1 M NaDBS in water	Potentiostatic 0.5-0.6 V vs. Ag/AgCl	10–30 min	RT	[33, 34]

Note: TBA PF_6: tetraethylammonium hexafluorophosphate.

In some applications, it is either impractical or undesirable to synthesize the polymer on a conducting surface. In these cases, chemical synthesis can be employed. As an example, the actuator as developed by Vidal et al. could be mentioned [29]. PEDOT is chemically synthesized in a poly(ethylene oxide)-polycarbonate (PEO-PC) network, building a PEDOT gradient toward the center, forming, in principle, a PEDOT/PEO-PC/PEDOT triple layer actuator. An additional fabrication option is solution casting, as has been used to fabricate PAni-based actuators [14, 28].

The choice of dopant species is of critical importance, determining both the actuation scheme [see Equation (8.1) and Equation (8.2)] and the amount of volume change. For instance, a study on the effect of the alkyl chain length in alkylbenzenesulfonate dopants showed that the medium sized octylbenzenesulfonate gave the largest strain [35]. By exchanging the dopant BF_4 for CF_3SO_3, the strain of PPy increased to 12% [36] and by switching to bis(trifluoromethanesulfonyl)imide as the dopant, the strain increased further to 26% [37]. Apart from these more exotic ions, typical dopant ions used are DBS [9, 38, 39], PF_6 [32, 40], BF_4 [41], and ClO_4 [19].

The solvent used during synthesis is another factor that determines the performance of the material. Most commonly, water [9], acetonitrile [25], or propylene carbonate [40] have been used. Bay et al. added pentanol as a co-surfactant to water and were able to increase the strain from 2.5% to 5.6% [42]. Likewise, Hara and coworkers showed that using methyl benzoate as the solvent resulted in a strain of up to 12% [36].

One limiting factor is the decreased electrical conductivity in PPy and other conjugated polymers in the reduced state. This results in a so-called IR drop in the material, that is, the potential decreases in the polymer with distance from the conducting (metal) contact, resulting in only some of the material being redox active. For example, Della Santa et al. observed that only the first 30 mm closest to the electrical contact of a 90-mm PPy strip was active during cycling [43]. In order to overcome this issue, a conducting material can be added in close contact with the conducting polymer. In bending beam actuators, the polymer is often synthesized on a conducting layer that later forms a part of the actuator and functions as a current collector [9, 44, 45]. Hutchison et al. platinized a PPy strip and obtained greater force from the strip used as a linear actuator [46]. The tube actuator as developed by the Wollongong group has a helical metal wire embedded into the polymer to assure good electrical conductivity throughout the entire actuator [32]. In addition, the wire supplies mechanical support and endurance. Similar approaches have been taken by others including electropolymerizing PPy on a microfabricated meander-type electrode and on a spring [47].

8.4 ACTIVATION AND CONTROL

Due to the nature of the actuating mechanism, an ion source/sink is needed. In laboratory settings, the actuator, a strip, or bending beam is often submerged into a liquid electrolyte similar to that used during electrochemical synthesis, but without the monomer. However, in order to get a more functional device, the actuator material, the electrolyte, and the counter electrode should be integrated into a single unit that can be operated under normal conditions. Several approaches have been taken.

Bending beams are often sandwiched as multi-layer devices comprising at least two electrodes on both sides of an electrolyte, and that may further include metallic layers for better electrical conduction or an encapsulation. The electrolyte may be a solid polymer electrolyte, a gel electrolyte, or a liquid electrolyte/ionic liquid soaked into a matrix such as paper or an ion exchange membrane [14, 29, 48–51]. Schemes involving an integrated electrolyte reservoir [32] or device encapsulation [40] have also been demonstrated. In biomedical applications, the surrounding fluids can be used as an electrolyte, making the devices less complicated because the electrolyte does not have to be integrated. Salt solutions, cell culture media, blood plasma and whole blood, urine, and x-ray contrast media have all been used as electrolytes [38].

In order to design an actuator unit, many parameters have to be taken into account because they all influence the performance of the unit. A systems approach should to be taken. The parameters include electrolyte (solvent type, composition, concentration, and pH) [23, 36, 52–54], electrodes (2 electrodes vs. 3 electrodes, position, and material) [23], signal (current vs. potential, shape, limits) [23, 26, 55–57], and temperature [36]. In addition, one must also take into account mechanical aspects such as number, thickness, and Young's modulus of the different layers of the active unit [24] and micropatterns to increase or direct the movement [58, 59]. Finally, adhesion between the different layers can be an issue, due to the large stresses that can occur between the various layers [26, 60, 61].

8.5 MACRO DEVICES

Development of conjugated polymer actuators has been primarily on linear actuation and bending beams. Strips [13], tubes [32], or fibers [46] as well as several bending beams such as those in Reference 9 have been made and characterized, but few devices have been fabricated.

Madden et al. have investigated the possibilities of CP actuators to deflect camber foils, that is, the tail sections of propeller blades [62]. Both bending beams, that is, bending the flaps directly, and linear strain actuation were evaluated. It was possible to actuate the foils on a small prototype, but scaling to full size will lead to issues with stacking of many thick layers and long actuation periods.

A Braille display has been developed by Ding et al. using the linear helical tube actuators mentioned previously [32]. The tube actuators, embedded in a single unit comprising an ionic liquid electrolyte, CE, and mechanical biasing spring, were able to drive a small pin up and down. Each pin formed a single "pixel" of the Braille display. A movement of approximately 0.2 to 0.6 mm was achieved, within the right order of magnitude for the intended application.

8.5 MICROACTUATORS

In designing devices, tradeoffs must be made between conflicting requirements such as force output (proportional to the amount of material, thus, requiring thick PPy) and speed requiring thin PPy, since ion transport into the polymer layer is generally rate limiting). Microfabrication is an appealing route to increase device performance while avoiding cumbersome material development. In addition, the large strains and bending angles demonstrated by conjugated polymers are unmatched by the conventional actuator materials used in microsystems or MEMS devices, for example piezoelectrics [63] and shape memory alloys [64]. Therefore, conjugated polymers are well suited to be employed as electro-active components in microsystems.

Using slightly adapted microfabrication methods, Smela et al. demonstrated the first conjugated polymer microactuators based on simple bending bilayers of Au and PPy(DBS) [44]. The complexity of the devices was rapidly increased by adding additional elements such as rigid plates [65], multiple actuators [45], and sensors [66], leading to increased functionality.

8.6 MICROFABRICATION

As mentioned previously, most of the technologies used in fabricating devices, comprising PPy actuators, are standard microfabrication methods such as photolithography. However, they have been adapted to fit conjugated polymers. Especially, two critical processes have been adapted: patterning of the conjugated polymer material and release of the actuator or device.

Patterning of conjugated polymers has been achieved using a wide range of techniques including lift-off [67], etching [68], and microcontact printing [69]. However, only a limited number of methods are generally used in the development of PPy microactuators. One method is synthesis in photoresist openings: PPy is electrochemically synthesized in a pattern of openings in a layer of photoresist applied on a conducting surface. Patterning can also be achieved by synthesis on patterned electrodes. In addition, PPy can be substractively patterned using reactive ion etching in an oxygen plasma. Even mechanical patterning methods have been developed. PPy and entire PPy-based devices have been cut (out) using laser ablation and punching [70].

Initial release of the actuator from the fabrication substrate is often an issue in microsystem development. When a release step is required, such as for bending beam PPy microactuators, several methods have been developed over the years. One of the most common methods of release utilizes a property called differential adhesion [65] (Figure 8.3a). This method is based on the poor adhesion between Au and Si (or SiO_2). An adhesive frame of Cr or Ti is patterned that surrounds the actuator, except for an anchoring point. An Au layer is deposited onto this patterned layer,. The Au functions as both the passive structural layer in the bilayer actuator and a current collector. Hereafter, further processing is performed including application of the patterned active PPy layer. Finally, the Au layer is etched, defining the lateral actuator dimensions. After the Au etching, the actuator is no longer attached to the adhesive frame. It is only attached to the substrate at the anchor point. Elsewhere, it is held onto the substrate only by the poor adhesive forces between the Si surface and Au layer. When the PPy microactuator is set in motion, it pulls itself free from the surface.

A limitation of the differential adhesion method is that it is restricted to a few combinations of materials, such as Au and Si. Therefore, other methods have been developed. Sacrificial layer and bulk etching are standard methods in microsystem fabrication to release devices, such as actuators, from a surface and these methods have been adapted for PPy actuators as well (Figure 8.3b and Figure 8.3c) [24, 45].

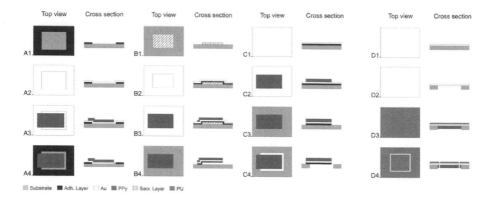

FIGURE 8.3 Schematic overview of different fabrication methods: (a) Differential adhesion, (b) sacrificial layer, (c) bulk etching, and (d) membrane fabrication. A1—Deposition and patterning of the adhesion layer (e.g., Cr) on the substrate (e.g., Si). A2—Deposition of Au layer. A3—Deposition and patterning of the PPy layer. A4—Patterning and etching of the final PPy microactuator structure. The PPy microactuator will be released by actuation due to the poor Au-Si adhesion. B1—Deposition and patterning of sacrificial layer (e.g., Ti) on the substrate. B2—Deposition of the adhesion and Au layers. B3—Deposition and patterning of the PPy layer. B4—Patterning and etching of the final PPy microactuator structure and etching of the sacrificial layer, resulting in a free hanging PPy microactuator. C1—Deposition of the adhesion and Au layers. C2—Deposition and patterning of the PPy layer. C3—Patterning and etching of the final PPy microactuator structure. C4—Bulk etching of the substrate in order to release the PPy microactuator. D1—Deposition of Au layer on substrate. D2—Bulk etching of the substrate, resulting in an Au membrane. D3—Deposition of PPy on the bottom side and polyurethane layer on the top side. D4—Cutting out the PPy actuator from the substrate frame.

This provides the opportunity to release structures that are more complex, such as individually actuated hinges or objects manufactured on Si wafers [71, 72].

All of the previous methods are based on bottom up fabrication, that is, layers are added successively onto the substrate thus building the actuator or device. However, certain applications (e.g., medical) require a fabrication method that gives access to both sides of the actuator during fabrication in order to minimize contamination with different processing solutions. This resulted in the development of a fourth fabrication path (Figure 8.3d) [70]. In this method, an Au membrane, the passive structural layer and current collector of the PPy actuator, is first made by bulk etching the substrate. Next, this membrane can be accessed from both sides to apply, for example, both the PPy active layer and a polyurethane blood-compatible polymer on opposing sides. Finally, the device can be cut out of the frame.

8.7   MICROSYSTEM DEVICES

8.7.1   Hinges

The basic actuator devices are PPy/Au bending bilayers [27, 44, 73]. These were then combined with other microfabricated elements such as rigid plates made of an inert, photopatternable polymer. The bending bilayers functioned as micro-hinges 30 µm × 30 µm in dimension [65]. The moving plates were used to create an active surface, the properties of which could be changed by flipping the 90 µm × 90 µm plates. In addition, multiple plates were connected together using multiple hinges. When the hinges were activated, the plates self-assembled into a cube of 300 µm × 300 µm × 300 µm. By reversing the applied potential, the cubes could be unfolded. The bending bilayers can also be used to grab micro-scale objects [38].

8.7.2   MICROROBOT

A more complex device that uses the PPy actuators as hinges is a microrobot. The microrobot was designed as an arm, having two hinges as an "elbow joint," two hinges as a "wrist," and three hinges as fingers of a "hand," all interconnected by rigid elements of SU8 [45]. The joints were individually controlled. Using this arm, a 100-µm bead could be moved over a surface (Figure 8.4). By actuating the two hinges of the elbow joint in counter phase, that is, bending one and stretching the other, the arm could even rotate somewhat.

In order to fabricate this complex device, a sacrificial layer release method was developed (Figure 8.3b). The robotic arm was 670-µm long and 250-µm wide. The robot elegantly illustrates the capabilities of the PPy actuator technology.

Using a similar design and fabrication, actuation in the same plane as the substrate was demonstrated [71]. Until that time, the bending bilayers could move only perpendicular to the substrate.

8.7.3   CELL CLINIC

A more application-based device is the so-called cell clinic [66]. This device was designed to perform biological studies on single cells. It consisted of a microvial with a lid actuated by PPy hinges to contain a cell or cells and an impedance sensor of two parallel Au microelectrodes on the bottom of the vial (Figure 8.5). The PPy actuator hinges, lid, and electrodes were all monolithically fabricated on a common substrate. A thick, 20-µm layer of pattern SU8 formed the microvial on top of the patterned electrodes. This eliminated the need for complex fabrication and pattern-ing techniques to form electrodes on the bottom of a microvial, which are generally made by etching cavities in an Si substrate [74, 75].

The lid was 150 × 150 µm^2, the hinges were 100 × 50 µm^2, the microvial was 100 × 100 µm^2, and the electrodes were 10 × 80 µm^2. The lid could be opened and closed by activating the PPy hinges. *Xenopus laevis* melanophore cells were seeded in the vials (and on the device). These cells have pigment granules that aggregate and disperse upon (bio-)chemical stimulation. Impedance measurements by means of the micro-electrodes on the bottom of the microvial were used to follow this intracellular event.

The cell clinic has paved the way for further cell-based sensing [76]. The same scheme of fabricating the vials as a thick rim on top of an Si-based sensor system is employed. The device will also include an on-chip potentiostat [77] and on-chip CE and RE [78].

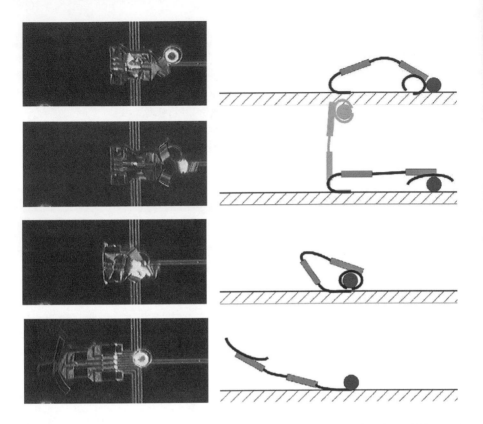

FIGURE 8.4 A sequence of frame-grabbed pictures (left column) showing the grabbing and lifting of a 100-μm glass bead by a PPy actuated microrobotic arm. The respective schematic illustrations of the sequence are shown in the right column. The robot arm has an "elbow," a "wrist," and *three* "fingers" placed 120° from each other. The pictures do not show the fact that the bead is actually lifted from the surface before it is placed at the base of the robot arm. We have illustrated this in grey in the second sketch to the right. (From Jager, E.W.H., O. Inganäs, and I. Lundström, Microrobots for micrometer-size objects in aqueous media: Potential tools for single cell manipulation. Science, 2000. 288(5475): 2335–2338. With permission.)

8.7.4 MICROVALVES

Microfluidics and drug delivery are another area where conjugated polymer micro-actuators may play a significant role. They have been demonstrated in a number of valve devices using different mechanical principles. Pettersson et al. have constructed a microvalve based on a moving plate [79]. A PPy hinge rotated a rigid plate inside a poly(dimethylsiloxane) (PDMS) microfluidic channel. By lifting the plate, the flow was to be stopped. A similar type of valve has been developed by Madou and coworkers for their drug delivery capsule. The initial idea of using the bulk volume change of PPy to open and close a cavity proved unfeasible [80] and was replaced by a hinge and plate design similar to the cell clinic [81].

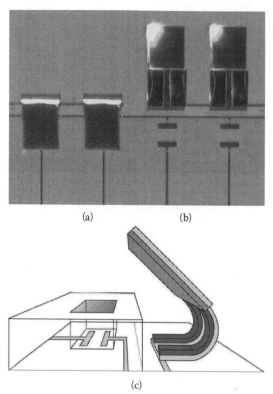

FIGURE 8.5   Two cell clinics in the (a) closed and (b) opened states. In the opened devices, two electrode pairs on the bottom of the transparent microvial are visible, as well as the PPy-microactuator hinges (black). (c) An illustration of the cell clinic. (From Jager, E.W.H. et al., The cell clinic: Closable microvials for single cell studies. Biomedical Microdevices, 2002. 4(3): 177–187. With permission.)

Another PPy valve was developed by HSG-IMIT and Micromuscle AB [82, 83]. This valve was developed to control the flow (on/off) of a drug delivery implant device called Intellidrug. The implant was the size of two molar teeth and was intended to temporally control the release of a drug into the mucosal tissue in the mouth. Due to the application and the small size of the device, power consumption of the valve was an important design criterion. PPy actuators fulfilled the specifications best with regard to power consumption, size, and driving voltage [83]. The valve was designed as a buckling membrane valve over an orifice and was constructed as a normally closed valve for safety reasons. In the first devices the in- and outlets, PPy/Au membrane, CE, and RE were all integrated into a single Si chip. However, the membrane did not close perfectly in the off state. This was due to the irreversible volume change of PPy in the first cycle. Therefore, in the second generation, a free-standing Au/PPy(DBS) bilayer membrane was made that was clamped between the fluidic circuitry comprising the orifice and the container that comprises the electrolyte and CE (Figure 8.6a). The freestanding membrane was precycled and clamped after the irreversible volume change. A separate electrolyte (0.1 M NaDBS) was used in contact with the "back side" of the membrane, that is, the PPy side, in order to separate the actuation from the liquid flow. Membrane deflection of $200 \pm 50\ \mu m$ was achieved. Figure 8.6b shows a flow profile of the valve. The on/off switching has a relatively large periodicity that fits the application.

An elegant use of the bulk volume change in a microvalve has been shown by Berdichevsky and Lo [84]. They mounted a microfluidic channel made in PDMS on top of an electrode covered with a thick layer of PPy. Upon activation, the volume of the PPy expanded and pushed against the thin bottom of the PDMS channel. The bottom then bulged upward, pinching off the flow.

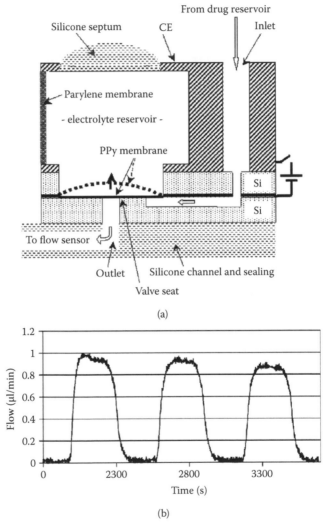

FIGURE 8.6 (a) A sketch of the buckling PPy membrane valve used in the Intellidrug device. (b) The flow profile of the valve. (From Göttsche, T. and S. Haeberle, Integrated oral drug delivery system with valve based on polypyrrole, in *Biomedical Applications of Electroactive Polymer Actuators*, F. Carpi and E. Smela, Editors. 2009, Wiley, New York. With permission.)

8.7.5 OTHER BULK EXPANSION DEVICES

Yet another device using the bulk volume change of PPy was a proposal for a device for tactile sensing. PPy was synthesized on comb-shaped microelectrodes and covered with a polyethylene glycol solid polymer electrolyte containing $LiClO_4$ as a salt. A 30 µm PPy layer expanded 2 µm in 30 s, which, according to the authors, should be enough to be sensed [85].

Wang and coworkers also used the bulk volume change to create a controlled wettability surface [86]. PPy was polymerized surrounding micropillars made of thick SU8, thus forming a PPy mesh. By activating the PPy mesh, the area between the pillars raised. This changed the surface from hydrophobic (a droplet was laying on the SU8 micropillars, fakir situation) to hydrophilic (the droplet was wetting the entire surface).

8.8   MEDICAL APPLICATIONS

Since the actuation mechanism requires the presence of an electrolyte, applications involving the liquid environments common to cell biology, biotechnology, and medicine seem evident. Indeed, some of these applications have been addressed previously in the discussion on microactuators.

One medical application that has seen significant progress is a PPy-based rotatable balloon seal developed in collaboration between Micromuscle AB and Boston Scientific. In bifurcation stenting, two stents are positioned around the bifurcation, one stent in the main artery and one in the side branch using a two-guide-wire system. The first stent has an opening through which the second stent is to be placed. It is therefore important that the first stent be positioned correctly, with the opening over the side branch. For this purpose, a rotating balloon system has been developed [87]. When inflating the dilation balloon, the pivot points at both ends of the balloon have to be sealed in order to mechanically lock the balloon in place so that it cannot rotate and in order to be able to build up a liquid pressure to unfold the balloon and deploy the stent. A PPy valve has been developed for this seal (Figure 8.7) [23, 88]. Inside a dilation balloon (number 216 in Figure 8.7), a metal ring is mounted on a balloon shaft (numbers 212 and 214) onto which a thick (40 to 70 µm) layer of PPy is electrosynthesized (numbers 230 and 232). The counter electrode (number 257) is mounted on the same shaft, inside the balloon. The PPy seal was tested in a dilation balloon test system using a standard physiological saline solution. Activation of the PPy resulted in expansion up to 30% of the PPy layer. This expansion was sufficient to close the gap between the shaft (number 230 or 232) and balloon pivot point (number 220 or 222), locking the balloon in place and preventing rotation. In addition, the PPy ring was able to seal the balloon from leaking and to withstand inflation pressures up to 24 atm, exceeding the clinically used pressures of 12 to 20 atm.

It was also shown that standard ethylene oxide sterilization, as used in the medical device industry, did not affect the performance of the PPy seal. After the initial proof of concept, the PPy seal was optimized with respect to maximum expansion and speed [23]. After careful consideration of the synthesis conditions, dopants, temperature, applied potential, and electrolyte concentration and composition, the maximum expansion was increased to 15 µm for a 50-µm thick PPy ring activated at −1.3

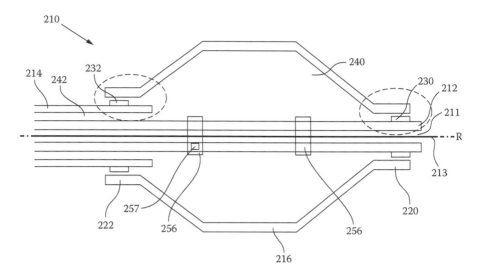

FIGURE 8.7 Sketch of the rotating balloon system, comprising a PPy seal (230, 232). (From Jager, E., D. Carlsson, M. Krogh, and M. Skoglund, Patent WO2008113372 Electroactive polymer actuator devices and systems comprising such devices, 2007.)

V in 0.3 M LiCl at a temperature of 37°C. The expansion speed increased as well: an expansion of 11 µm was achieved in the first few seconds.

Another medical device developed by Micromuscle AB is a microanastomosis connector [89]. The device consists of a rolled-up tube, made of a triple-layered sheet of blood-compatible polyurethane, Au, and PPy and further includes an Au micropattern that predetermines the bending direction [59]. It is intended to reconnect two ends of a small, 1 to 3 mm diameter blood vessel, divided either by trauma or by a surgical procedure, as an alternative to extensive suturing (Figure 8.8). By using a small potential the tube diameter is reduced by contracting, that is, rolling up the sheet more tightly. The tube is then inserted in both ends of the divided vessel and the potential is disconnected. The tube diameter expands by unrolling the sheet and holds the two vessel ends together while the vessel can heal. The materials and device have passed basic biocompatibility testing, including cytotoxicity, irritation, acute systemic toxicity, and hemolysis. A small implantation study of the connector in a rat model has also been performed. After three months, the devices showed not to be causing blood clotting or other obstruction, indicating blood- and biocompatibility. Coagulation studies using a Chandler looping setup were also performed on both heparinized (in order to increase blood compatibility) and untreated, as fabricated PPy connectors. These studies indicated that there was little difference in coagulation between both the heparinized and unheparinized PPy connectors and commercial stents.

Conjugated polymer actuation has often been proposed as a mechanism to bend guide wires and catheters in order to increase the ease with which such devices are guided through the vascular system. One common design strategy has been to add strips or bending bilayers on four opposing sides of the tubular device. In this way

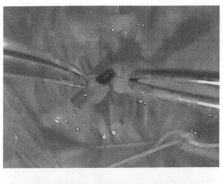

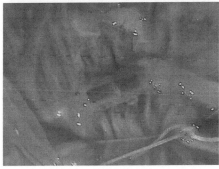

FIGURE 8.8 Ex vivo surgery using a microanastomosis connector. The microanastomosis connector is an implantable tubular device that is intended for reconnecting millimeter-sized blood vessels. (From Wilson, S.A. et al., New materials for micro-scale sensors and actuators. An engineering review. Materials Science & Engineering R-Reports, 2007. 56(1-6): 1–129. With permission.)

bending in both x and y direction, that is, a 360° circular motion of the tip, should be achieved. This strategy was used by Lee et al. [90]. They modified a standard 0.5-mm OD catheter by applying one or two opposing pairs of PPy electrodes on the tip. Upon activation in a (non-physiological) 1 M $NaPF_6$ solution, bending curvatures up to 0.06 m^{-1} were achieved. An alternative design concept was presented by Micromuscle AB [91]. PPy is applied on one side of the coil that makes up the tip of a guide wire (Figure 8.9). Due to asymmetric application of PPy, a volume expansion results in a bending motion of the tip. A full 360° circular motion of the tip can be achieved by rotating the guide wire or catheter shaft, as is common clinical practice using standard fixed curvature guide wires. This design makes electronic control and device fabrication extremely easy. Only one pair of electrodes (one of which is the PPy actuator) must be controlled. Thus, only two electrical leads have to be integrated into the guide wire. Figure 8.9 shows how such a device can be maneuvered through a mock-up vascular system by electroactively changing the tip curvature. The surrounding fluid, in this case a salt solution, was used as the electrolyte.

The same issue of changing the curvature during a surgical procedure exists for cochlear implants. It would be advantageous if the curvature of electrodes for cochlear implants could be altered actively during the insertion procedure.

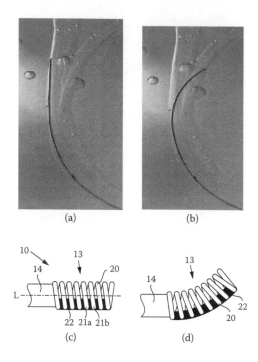

FIGURE 8.9   A PPy-activated guide wire for maneuvering in vessels. (a, b) Demonstrated in a bench top set-up and (c, d) schematic drawing of the principle. (a, c) PPy is in the contracted state and the guide wire is straight. (b, d) PPy is in the expanded state and the guide wire tip bends. (a, b from Wilson, S.A. et al., New materials for micro-scale sensors and actuators. An engineering review. Materials Science & Engineering R-Reports, 2007. 56(1-6): 1–129. With permission. c, d from Krogh, M. and E. Jager, Patent WO2007057132, Medical Devices and Methods for Their Fabrication and Use, 2005.)

Therefore, a cochlear electrode array that can electroactively bend using PPy has been developed by the University of Wollongong [92]. The active element was a PPy/Pt/PVDF/Pt/PPy multilayer actuator, with the PVDF pores filled with a propylene carbonate-based electrolyte. The multilayer actuator was mounted on the back of a standard cochlear electrode array. Bending of almost 180° could be achieved.

8.9   CONCLUSION

Since the first demonstration of conjugated polymer actuators, or artificial muscles, in the early 1990s, the technology has made huge progress, both in understanding the properties and improving performance such as strains and stresses. Several sophisticated device demonstrators and prototypes have been demonstrated. Development of the conjugated polymer actuator technology still continues and will pave the way for novel (commercial) applications.

8.10   ACKNOWLEDGMENTS

The author wishes to thank Dr. Daniel Simon for his help and valuable feedback on the manuscript. The author has been chief technology officer of Micromuscle AB, Linköping, Sweden from 2000 to 2007 and wishes to thank his former colleague Mr. Magnus Krogh for his feedback on the manuscript.

REFERENCES

1. Chiang, C.K. et al., Electrical conductivity in doped polyacetylene. Physical Review Letters, 1977. 39(17): 098–1101.
2. Foot, P.J.S. and R. Simon, Electrochromic properties of conducting polyanilines. Journal of Physics D: Applied Physics, 1989. 22(11): 1598–1603.
3. Isaksson, J. et al., Solid-state organic electronic wettability switch. Advanced Materials, 2004. 16(4): p. 316–320.
4. Wang, X., T. Ederth, and O. Inganäs, In situ wilhelmy balance surface energy determination of poly(3-hexylthiophene) and poly(3,4-ethylenedioxythiophene) during electrochemical doping-dedoping. Langmuir, 2006. 22(22): 9287–9294.
5. Miller, L.L., Electrochemically controlled release of drugs and other chemicals. Reactive Polymers, 1987. 6(2–3): 341–341.
6. Pyo, M. and J.R. Reynolds, Electrochemically stimulated adenosine 5'-triphosphate (ATP) release through redox switching of conducting polypyrrole films and bilayers. Chemistry of Materials, 1996. 8(1): 128–133.
7. Wadhwa, R., C.F. Lagenaur, and X.T. Cui, Electrochemically controlled release of dexamethasone from conducting polymer polypyrrole coated electrode. Journal of Controlled Release, 2006. 110(3): 531–541.
8. Baughman, R.H. et al., Micro electromechanical actuators based on conducting polymers, in *Molecular Electronics*, P.I. Lazarev, Editor. 1991, Kluwer Academic Publishers, Dordrecht. 267–289.
9. Pei, Q. and O. Inganäs, Conjugated polymers and the bending cantilever method: electrical muscles and smart devices. Advanced materials, 1992. 4(4): 277–278.
10. Otero, T.F., E. Angulo, J. Rodriguez, and C. Santamaria, Electrochemomechanical properties from a bilayer—polypyrrole nonconducting and flexible material artificial muscle. Journal of Electroanalytical Chemistry, 1992. 341(1–2): 369–375.
11. Gandhi, M.R., P. Murray, G.M. Spinks, and G.G. Wallace, Mechanism of electromechanical actuation in polypyrrole. Synthetic Metals, 1995. 73: 247–256.
12. Baughman, R.H., Conducting polymer artificial muscles. Synthetic Metals, 1996. 78: p. 339–353.
13. Della Santa, A., D. De Rossi, and A. Mazzoldi, Characterization and modelling of a conducting polymer muscle-like linear actuator. Smart Material Structure, 1997. 6: 23–34.
14. Kaneto, K., M. Kaneko, Y. Min, and A.G. MacDiarmid, 'Artificial muscle': Electromechanical actuators using polyaniline films. Synthetic Metals, 1995. 71(1-3): 2211–2212.
15. Pei, Q. and O. Inganäs, Electrochemical applications of the bending beam method. 1. Mass transport and volume changes in polypyrrole during redox. Journal of Physical Chemistry, 1992. 96(25): 10507–10514.
16. Pei, Q. and O. Inganäs, Electrochemical applications of the bending beam method. 2. Electroshrinking and slow relaxation in polypyrrole. Journal of Physical Chemistry, 1993. 97(22): 6034–6041.

17. Pei, Q. and O. Inganäs, Electrochemical applications of the bending beam method: A novel way to study ion transport in electroactive polymers. Solid State Ionics, 1993. 60: 161–166.

18. Bay, L., T. Jacobsen, S. Skaarup, and K. West, Mechanism of actuation in conducting polymers: Osmotic expansion. Journal of Physical Chemistry B, 2001. 105: 8492–8497.

19. Otero, T.F., H. Grande, and J. Rodriguez, Conformational relaxation during oxidation: from experiment to theory. Electrochimica Acta, 1996. 41(11–12): 1863–1869.

20. Otero, T.F. and J. Padilla, Anodic shrinking and compaction of polypyrrole blend - electrochemical reduction under conformational relaxation kinetic control. Journal of Elctroanalytical Chemistry, 2004. 561: 167–171.

21. Smela, E. and N. Gadegaard, Surprising volume change in PPy(DBS): An atomic force microscopy study. Advanced Materials, 1999. 11(11): 953–957.

22. Smela, E. and N. Gadegaard, Volume change in polypyrrole studied by atomic force microscopy. Journal of Physical Chemistry B, 2001. 105: 9395–9405.

23. Carlsson, D., E. Jager, M. Krogh, and M. Skoglund, Patent WO2009038501, Systems, device and object comprising electroactive polymer material, methods and uses relating to operation and provision thereof, 2007.

24. Smela, E., M. Kallenbach, and J. Holdenried, Electrochemically driven polypyrrole bilayers for moving and positioning bulk micromachined silicon plates. Journal of Microelectromechanical Systems, 1999. 8(4): 373–383.

25. Otero, T.F. and J.M. Sansiñena, Bilayer dimensions and movement in artificial muscles. Bioelectrochemistry and Bioenergetics, 1997. 42: 117–122.

26. Maw, S. et al., The effects of varying deposition current density on bending behaviour in PPy(DBS)-actuated bending beams. Sensors and Actuators A, 2001. 89: 175–184.

27. Zhou, J.W.L. et al., Polymer MEMS actuators for underwater micromanipulation. IEEE/ASME Transactions on Mechatronics, 2004. 9(2): 334–342.

28. Pei, Q., O. Inganäs, and I. Lundström, Bending bilayer strips built from polyaniline for artificial electrochemical muscles. Smart Material Structures, 1993. 2: 1–6.

29. Vidal, F. et al., Feasibility of conducting semi-interpenetrating networks based on a poly(ethylene oxide) network and poly(3,4-ethylenedioxythiophene) in actuator design. Journal of Applied Polymer Science, 2003. 90(13): 3569–3577.

30. Bohn, C., S. Sadki, A.B. Brennan, and J.R. Reynolds, In situ electrochemical strain gage monitoring of actuation in conducting polymers. Journal of the Electrochemical Society, 2002. 149(8): p. E281–E285.

31. Otero, T.F. and M.T. Cortes. Electrochemical characterization and control triple-layer muscles. In *Smart Structures and Materials 2000*. 2000. Proceedings of SPIE, Newport Beach, CA.

32. Ding, J. et al., High performance conducting polymer actuators utilising a tubular geometry and helical wire interconnects. Synthetic Metals, 2003. 138(3): 391–398.

33. Smela, E., Microfabrication of PPy microactuators and other conjugated polymer devices. Journal of Micromechanical Microengineering, 1999. 9(1): 1–18.

34. Wilson, S.A. et al., New materials for micro-scale sensors and actuators—An engineering review. Materials Science & Engineering R-Reports, 2007. 56(1–6): 1–129.

35. Bay, L. et al., Polypyrrole doped with aklyl benzenesulfonates. Macromolecules, 2002. 35: 9345–9351.

36. Hara, S., T. Zama, W. Takashima, and K. Kaneto, Artificial muscles based on polypyrrole actuators with large strain and stress induced electrically. Polymer Journal, 2004. 36(2): p. 151–161.

37. Hara, S., T. Zama, W. Takashima, and K. Kaneto, TFSI-doped polypyrrole actuator with 26% strain. Journal of Material Chemistry, 2004. 14: 1516–1517.

38. Jager, E.W.H., E. Smela, and O. Inganäs, Microfabricating conjugated polymer actuators. Science, 2000. 290(5496): 1540–1545.
39. Smela, E., Conjugated polymer actuators for biomedical applications. Advanced Materials, 2003. 15(6): 481–494.
40. Madden, J. et al., Encapsulated polypyrrole actuators. Synthetic Metals, 1999. 105: 61–64.
41. Hara, S. et al., Highly stretchable and powerful polypyrrole linear actuators. Chemistry Letters, 2003. 32(7): 576.
42. Bay, L., K. West, and S. Skaarup, Pentanol as co-surfactant in polypyrrole actuators. Polymer, 2002. 43(12): 3527–3532.
43. Della Santa, A., D.D. Rossi, and A. Mazzoldi, Performance and work capacity of a PPy conducting polymer linear actuator. Synthetic Metals, 1997. 90: 93–100.
44. Smela, E., O. Inganäs, Q. Pei, and I. Lundström, Electrochemical muscles: Micromachining fingers and corkscrews. Advanced Materials, 1993. 5(9): 630–632.
45. Jager, E.W.H., O. Inganäs, and I. Lundström, Microrobots for micrometer-size objects in aqueous media: Potential tools for single cell manipulation. Science, 2000. 288(5475): 2335–2338.
46. Hutchison, A.S. et al., Development of polypyrrole-based electromechanical actuators. Synthetic Metals, 2000. 113(1–2): 121–127.
47. Hara, S. et al., Polypyrrole-metal coil composites as fibrous artificial muscles. Chemistry Letters, 2003. 32(9): 800–801.
48. Lewis, T.W. et al., Development an all polymer electromechanical actuators. Polymer Reprints, 1997. 38: 520–521.
49. Lu, W. et al., Use of ionic liquids for -conjugated polymer electrochemical devices. Science, 2002. 297(5583): 983–987.
50. Sansiñena, J.M. et al., A solid state artificial muscle based on polypyrrole and a solid polymeric electrolyte working in air. Chemical Communications, 1997. 22: 2217–2218.
51. Wu, Y., G. Alici, G.M. Spinks, and G.G. Wallace, Fast trilayer polypyrrole bending actuators for high speed applications. Synthetic Metals, 2006. 156(16–17): 1017–1022.
52. Shimoda, S. and E. Smela, The effect of pH on polymerization and volume change in PPy(DBS). Electrochimica Acta, 1998. 44: 219–238.
53. Nakashima, T. et al., Enhanced electrochemical strain in polypyrrole films. Current Applied Physics, 2005. 5: 202–208.
54. Maw, S., E. Smela, K. Yoshida, and R.B. Stein, Effects of monomer and electrolyte concentrations on actuation of PPy(DBS) bilayers. Synthetic Metals, 2005. 155(1): 18–26.
55. Spinks, G.M. et al., Enhanced control and stability of polypyrrole electromechanical actuators. Synthetic Metals, 2004. 140(2–3): 273–280.
56. Lewis, T.W. et al., Investigation of the applied potential limits for polypyrrole when employed as the active components of a two-electrode device. Synthetic Metals, 2001. 122(2):379–385.
57. Madden, J.D., R.A. Cush, T.S. Kanigan, and I.W. Hunter, Fast contracting polypyrrole actuators. Synthetic Metals, 2000. 113(1–2): 185–192.
58. Bay, L. et al., A conducting polymer artificial muscle with 12% linear strain. Advanced Materials, 2003. 15(3): 310–313.
59. Krogh, M., O. Inganäs, and E. Jager, Patent WO03039859, Fibre-reinforced microactuator, 2001.
60. Bohn, C.C. et al., In-situ strain measurements of polypyrrole actuators on enhanced Au surfaces. Polymeric Materials: Science & Engineering, 2002. 86: 26–27.
61. Krogh, M., Patent WO2008121033, Use of a material in a device, a device, and applications and a method for fabrication thereof, 2007.

62. Madden, J., Application of EAPs: Polypyrrole variable camber propellers. In *Smart Structures and Materials; Electoactive Polymer Actuators and Devices (EAPAD).* 2004. SPIE, San Diego, CA.

63. DeVoe, D.L. and A.P. Pisano, Modeling and optimal design of piezoelectric cantilever microactuators. Journal of Microelectromechanical Systems, 1997. 6(3): 266–270.

64. Krulevitch, P. et al., Thin film shape memory alloy microactuators. Journal of Microelectromechanical Systems, 1996. 5(4): 270–282.

65. Smela, E., O. Inganäs, and I. Lundström, Controlled folding of micrometer-size structures. Science, 1995. 268: 1735–1738.

66. Jager, E.W.H. et al., The cell clinic: Closable microvials for single cell studies. Biomedical Microdevices, 2002. 4(3): 177–187.

67. Chan, J.R., X.Q. Huang, and A.M. Song, Nondestructive photolithography of conducting polymer structures. Journal of Applied Physics, 2006. 99(2): Art. No. 023710, 1–4.

68. Balocco, C., L.A. Majewski, and A.M. Song, Non-destructive patterning of conducting-polymer devices using subtractive photolithography. Organic Electronics, 2006. 7(6): 500–507.

69. Granlund, T. et al., Patterning of polymer light emitting diodes with soft lithography. Advanced Materials, 2000. 12(4): 269–273.

70. Jager, E. and M. Krogh, Patent WO2004092050, Method for producing a micromachined layered device, 2003.

71. Jager, E.W.H., O. Inganäs, and I. Lundström, Perpendicular actuation with individually controlled polymer microactuators. Advanced Materials, 2001. 13(1): 76–79.

72. Smela, E., A microfabricated movable electrochromic "pixel" based on polypyrrole. Advanced Materials, 1999. 11(16): 1343–1345.

73. Lee, A.P., K.C. Hong, J. Trevino, and M.A. Northrop, Thin film conductive polymer for microactuator and micromuscle applications. In *Dynamic and Systems and Control Session, International Mechanical Engineering Congress.* 1994. ASME Publications, Chicago, IL.

74. Jansson, M. et al., Micro vials on a silicon wafer for sample introduction in capillary electrophoresis. Journal of Chromatography, 1992. 626: 310–314.

75. Noort, D.v., J. Rumberg, E.W.H. Jager, and C.F. Mandenius, Silicon based affinity biochips viewed with imaging ellipsometry. Measurement Science & Technology, 2000. 11(6): 801–808.

76. Urdaneta, M. et al. Integrating conjugated polymer microactuators with cmos sensing circuitry for studying living cells. In SPIE Smart Structures and Materials; Electoactive Polymer Actuators and Devices (EAPAD). San Diego, CA, March 7–10, 2005. *Proc of SPIE* 5759: 232–240.

77. Prakash, S.B. et al., In situ electrochemical control of electroactive polymer films on a CMOS chip. Sensors and Actuators, B: Chemical, 2008. 129(2): 699–704.

78. Jager, E.W.H., E. Smela, and O. Inganäs, On-chip microelectrodes for electrochemistry with moveable PPy bilayer actuators as working electrodes. Sensors & Actuators B: Chemical, 1999. 56(1-2): 73–78.

79. Pettersson, F., E.W.H. Jager, and O. Inganäs. Surface micromachined polymer actuators as valves in pdms microfluidic system. In IEEE-EMBS Special Topic Conference on Microtechnologies in Medicine & Biology. 2000. Lyon, France.

80. Low, L.-M., S. Seetharaman, K.-Q. He, and M.J. Madou, Microactuators toward microvalves for responsive controlled drug delivery. Sensors and Actuators B: Chemical, 2000. 67: 149–160.

81. Xu, H. et al., Polymer actuator valves toward controlled drug delivery application. Biosensors and Bioelectronics, 2006. 21(11): 2094–2099.

82. Göttsche, T. et al., Patent DE102006005517 Ventil, 2006.

83. Göttsche, T. and S. Haeberle, Integrated oral drug delivery system with valve based on polypyrrole, in *Biomedical Applications of Electroactive Polymer Actuators*, F. Carpi and E. Smela, Editors. 2009, Wiley, New York.

84. Berdichevsky, Y. and Y.-H. Lo. Polymer microvalve based on anisotropic expansion of polypyrrole. In Mat. Res. Soc. Symp. Proc. 2004: Materials Research Society.

85. Yamada, K., Y. Kume, and H. Tabe, A solid-state electrochemical device using poly(pyrrole) as micro-actuator. Japanese Journal of Applied Physics Part 1-Regular Papers Short Notes & Review Papers, 1998. 37(10): 5798–5799.

86. Wang, X., M. Berggren, and O. Inganäs, Dynamic control of surface energy and topography of microstructured conducting polymer films. Langmuir, 2008. 24(11): 5942–5948.

87. Gumm, D., Patent WO03/017872, Rotating stent delivery system for side branch access and protection and method of using same, 2002.

88. Jager, E., D. Carlsson, M. Krogh, and M. Skoglund, Patent WO2008113372 Electroactive polymer actuator devices and systems comprising such devices, 2007.

89. Immerstrand, C. et al., Conjugated-polymer micro- and milliactuators for biological applications. MRS bulletin, 2002. 27(6): 461–464.

90. Lee, K.K.C. et al., Fabrication and characterization of laser-micromachined polypyrrole-based artificial muscle actuated catheters. Sensors and Actuators A: Physical, 2009. 153(2): 230–236.

91. Krogh, M. and E. Jager, Patent WO2007057132, Medical devices and methods for their fabrication and use, 2005.

92. Zhou, D. et al., Actuators for the cochlear implant. Synthetic Metals, 2003. 135(1-3): 39–40.

9 Organic Electrochemical Transistors for Sensor Applications

Sang Yoon Yang, Fabio Cicoira,
Nayoung Shim, and George G. Malliaras

CONTENTS

9.1 INTRODUCTION

Organic thin film transistors (OTFTs) are attracting a great deal of interest due to their potential for applications in flexible displays, RF-ID tags, and sensors.[1,2] OTFTs are a three-electrode device. The source and drain electrodes are in contact with the organic semiconductor channel (they supply the source-drain voltage and collect the source-drain current) and the gate electrode switches the channel between the "on" (high conductivity) and the "off" (low conductivity) states and modulates the magnitude of the source-drain current. This switching in OTFTs can be induced either field-effect or electrochemical doping, depending on what kind of layer is inserted between the organic semiconductor and the gate electrode. In organic field-effect transistors (OFETs), the semiconducting layer is separated from the gate electrode by the gate dielectric (Figure 9.1a). The density of charge carriers in the organic semiconductor channel is controlled by the application of potential to gate electrode, which results in a constant electric field (a linear electrostatic potential drop) across the gate dielectric.[3] A high operating voltage (tens

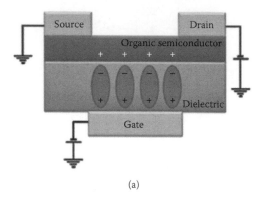

(a)

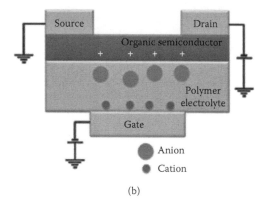

(b)

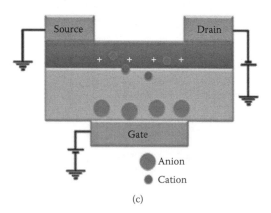

(c)

FIGURE 9.1 Device structures of an OFET with (a) a conventional gate dielectric (b) an OFET with electrolyte, and (c) an OECT with electrolyte. **(Continued)**

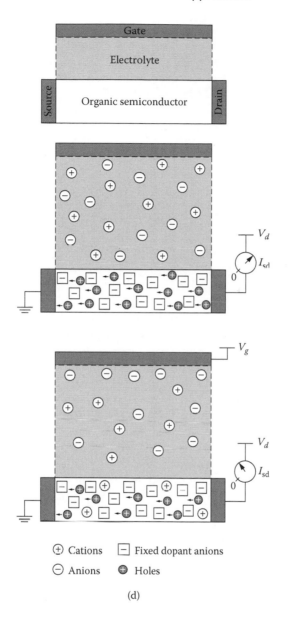

+ Cations   − Fixed dopant anions

− Anions   ● Holes

(d)

FIGURE 9.1 (Continued) (d) working principle of the OECT. (From Bernards, D.A. and Malliaras, G.G. Steady-state and transient behavior of organic electrochemical transistors. *Adv. Funct. Mater.* **17**, 3538–3544 (2007). With permission.)

of volts) needs to be applied in conventional OFETs because of the low capacitance of conventional gate dielectrics (usually less than 20 nF/cm^2).[4]

An alternative approach that allows a low operating voltage OFET involves the replacement of the gate dielectric with an electronically insulating, but ionically conducting solid electrolyte, as shown in Figure 9.1b.[5–9] When a gate bias is applied, two electric double layer capacitors (EDLCs) are established in series, located at the gate/electrolyte and the electrolyte/organic semiconductor interfaces. The electrolyte between them remains electrically neutral. The EDLCs have extraordinary high capacitance (10 ~ 500 μF/cm^2)[10] and, therefore, low-voltage operation (< 2V) accompanied by high field-induced charge-carrier density (10^{14} ~ 10^{15} charges/cm^2) in organic semiconductor can be achieved.[5–7,11–13] However, if the ions in electrolyte can penetrate the semiconductor active layer, an electrochemical doping process begins to govern the operation of the transistor (Figure 9.1c). The latter devices are called organic electrochemical transistors (OECTs) and incorporate the electrolyte as an integral part of their device structure.

OECTs consist of a channel usually made from a conducting polymer (i.e., a degenerately doped p-type organic semiconductor) in direct contact with an electrolyte.[14–17] A gate electrode immersed in the electrolyte controls the level of doping in the conducting polymer channel: When a positive bias is applied at the gate electrode, cations from the electrolyte enter the polymer film and de-dope it, decreasing its conductivity (Figure 9.1d). This process is reversible and occurs at gate and drain voltages that are below the threshold of electrolysis. If a highly conducting polymer is employed, the channel and the electrodes can be made by the same material, which considerably simplifies device fabrication. A convenient feature of OECTs, associated with the use of an electrolyte, is that the gate electrode does not need to be positioned at a small distance from the channel. This allows planar device architectures in which the area, shape, and relative position of the channel and the gate electrode can be varied independently. Therefore, the resulting low-voltage operation, the simplified device structure, and the ability to operate in aqueous environments enable OECT to be the essential transducer element for biological sensing applications.[15,17–21]

The first OECT was reported by White et al. in 1984, and utilized a potentiostat to control the conductivity of a polypyrrole layer, as shown in Figure 9.2.[22] In addition to polypyrrole, many other conjugate polymers such as polyaniline,[23–29] polycarbazole,[30] polythiophene, and their derivatives[31,32] have been used to fabricate OECTs for chemical and biological sensing. However, many conducting polymers have limitations when it comes to biosensor applications. For example, the conductivity of polypyrrole is irreversibly destroyed upon exposure to hydrogen peroxide (H_2O_2), limiting its use with enzymes such as glucose oxidase (GOx) that generate H_2O_2 during interaction with suitable analytes.[17,33] Moreover, polyaniline loses its electrochemical activity at a pH higher than 5, limiting the sensing capability of polyaniline-based OECTs in physiological fluid (pH ~ 7.3).[25] Although there have been successful attempts to overcome this limitation by modifying polyaniline with high molecular counter ions such as poly(vinyl sulfonate) or poly (styrene sulfonate),[25,34] the development of more robust conjugated polymers is still required to operate OECTs over a broad pH range.

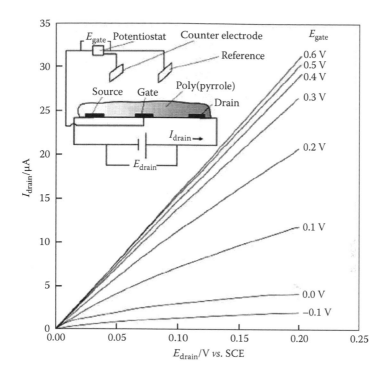

FIGURE 9.2 Current-voltage characteristics of the first OECT. Each curve corresponds to a different gate voltage. The inset shows a schematic of the device structure. (From White, H.S., Kittlesen, G.P., and Wrighton, M.S. Chemical derivatization of an array of 3 gold microelectrodes with polypyrrole—Fabrication of a molecule-based transistor. *J. Am. Chem. Soc.* **106**, 5375–5377 (1984). With permission.)

As shown in Figure 9.3a, Zhu et al. have demonstrated that OECTs based on the commercially available conducting polymer, poly(3,4-ethylenedioxythiophene) doped with poly(styrene sulfonic acid) (PEDOT:PSS), is capable of sensing glucose in a neutral pH buffer solution.[15] In this system, the current modulation in the PEDOT:PSS channel, induced by the application of a gate voltage on a platinum (Pt) wire electrode, was dramatically increased when both glucose and GOx were present in phosphate buffer solution (PBS). It was found that this large gating effect results from the production of H_2O_2 during the redox reaction cycle of glucose and GOx. This result indicates that PEDOT:PSS has good stability both in neutral pH and in the presence of H_2O_2 and the limitations that result from the use of polypyrrole and polyaniline can be overcome.

PEDOT:PSS is synthesized through the oxidative polymerization of ethylene dioxythiophene (EDOT) monomer with water-dispersible polyelectrolyte, PSS, as a charge-balancing counter-ion. This renders the resulting PEDOT:PSS dispersible in aqueous environment and therefore PEDOT:PSS is available as an aqueous dispersion. The conjugated polymer PEDOT is doped p-type and the anionic sulfonate group of PSS maintains charge neutrality (Figure 9.3b). PEDOT:PSS exhibit a high electrical conductivity (up to 10 S/cm), high transparency for visible light, and

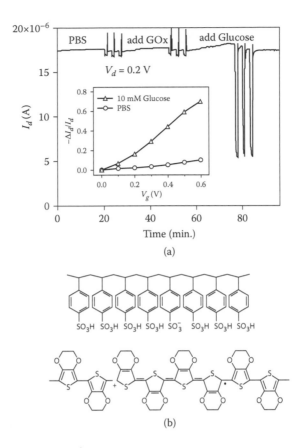

FIGURE 9.3 (a) Plot of drain current vs. time for a PEDOT:PSS OECT in PBS solution, where GOx was added first and then glucose was added. The gate voltage was pulsed from 0 to 0.6 V with 1 min of time interval. The inset shows the modulation of drain current as a function of gate voltage for solutions containing different concentrations of glucose. (From Zhu, Z.T. et al. A simple poly(3,4-ethylene dioxythiophene)/poly(styrene sulfonic acid) transistor for glucose sensing at neutral pH. *Chem. Commun.*, 1556–1557 (2004). With permission.) (b) Chemical structure of PEDOT:PSS. A positive polaron is indicated on the polythiophene backbone.

excellent chemical and environmental stability. PEDOT:PSS can be easily deposited to form a thin film on various substrates such as glass, plastic, and paper by spin-coating or printing techniques.[35–38] The conductivity of PEDOT:PSS can be tuned by the addition of polar solvents such as sorbitol, ethylene glycol, dimethyl sulfoxide, and *N,N*-dimethylformamide.[39–42] The effect is believed to originate from morphological changes in the PEDOT:PSS film and a conformational change of the PEDOT chains.[40,42]

Because of these advantages of PEDOT:PSS, OECTs based on this material have been studied for applications in biosensors and bioelectronics.[15,16,18,20,21,43–46] OECTs serve as converters between ionic currents in an electrolyte and electronic currents in the conducting polymer active layer (see later).[16,44] A sensor that detects ion valence was built based on this principle, by integration of an OECT with a bilayer lipid membrane in which the ion channel gramicidin was incorporated.[18] As gramicidin is only permeable to monovalent cations, gating of these OECTs occurred only when monovalent cations such as K^+ were present in the electrolyte. Ion selective OECTs were also demonstrated using a Ca^{2+} selective polymer membrane coated on a PEDOT:PSS channel.[46] In addition, OECTs can be operated in the Faradaic regime, where redox reactions at the gate electrode/electrolyte interface change the current in the polymer channel (see later).[21] Glucose sensors were developed based on this principle, by incorporating the enzyme GOx in the electrolyte.[15,20,21] PEDOT:PSS is also biocompatible, meaning that cells are able to adhere and proliferate on its surface. The biocompatibility of PEDOT:PSS makes it possible to develop novel devices such as ion pumps that mediate ion homeostasis of individual neuronal cells.[45] Cell adhesion and cell density also can be controlled electronically by switching the redox state of PEDOT-based conducting polymer films.[43,45]

9.2 MECHANISM OF OECT OPERATION

9.2.1 OECTs as Ion-to-Electron Converters

OECT-based sensors can be operated in the non-Faradaic and the Faradaic regimes, the former characterized by the absence of steady-state current flow through the gate circuit. The two regimes offer different opportunities in sensor applications. In non-Faradaic operation, the OECT acts as an ion-to-electron converter. In the Faradaic regime, on the other hand, a redox reaction alters the potential within the electrolyte and results in a change of the source-drain current (I_{sd}).

A simple model of operation for p-type OECTs in the non-Faradaic regime has been formulated by Bernards et al.[44] Typical electrical characteristics for a PEDOT:PSS OECT are shown in Figure 9.4a. As a convention, the source is grounded and a voltage relative to the ground is applied to the drain electrode (V_d). The current passing through the channel is monitored as a function of the gate voltage (V_g). At $V_g = 0$ V, the transistor is in its "on" state and a high current passes through the channel. Upon application of a positive V_g, cations from the electrolyte enter the organic semiconductor.[14] The reaction results in de-doping of the channel, which decreases I_{sd}. At low V_d, I_{sd} depends linearly on V_d, whereas at higher V_d it tends toward saturation (third quadrant, Figure 9.4a).

To simplify the modeling of these characteristics, the OECT can be divided into an electronic and an ionic circuit. The ionic circuit, which accounts for transport of ionic charge in the electrolyte, is described as a combination of linear circuit elements (Figure 9.4b). The electronic circuit, which consists of the p-type organic semiconductor film that transports holes between the source and drain electrodes, is described by Ohm's law:

$$J(x) = q\mu p(x)\frac{dV(x)}{dx}$$

$$(9.1)$$

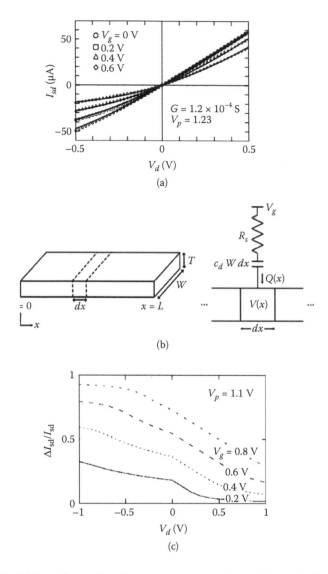

FIGURE 9.4 (a) Experimental steady-state current vs. voltage characteristics (data points) and fit to the model (solid lines) for $G = 1.2 \times 10^{-4}$ S and $V_p = 1.23$ V. A 10-mM NaCl solution was used as the electrolyte. The device had channel length (L) and channel width (W) of 5 mm and 6 mm. (b) Device geometry used in the OECT model. *(left)* Organic semiconductor film where the source is located at $x = 0$ and the drain at $x = L$; *(right)* charge (Q) from the ionic circuit is coupled to the voltage in the electronic circuit at a position x along the organic semiconductor. (c) Simulated steady-state device response as a function of drain voltage for a series of gate voltages with $V_p = 1.1$ V. I_{sd} is the drain current without an applied gate voltage and ΔI_{sd} is the change in drain current upon the application of a gate voltage. (From Bernards, D.A. and Malliaras, G.G. Steady-state and transient behavior of organic electrochemical transistors. *Adv. Funct. Mater.* **17**, 3538-3544 (2007). With permission.)

where J is the current density, q is the elementary charge, μ is the hole mobility, p is the hole density, and dV/dx is the electric field. Electronic transport in this circuit depends on hole density and mobility. Upon the application of a gate voltage, cations from the electrolyte enter the semiconductor film and compensate the PSS acceptors. For each compensating cation, a hole extracted at the source is not replaced by injection at the drain (assuming $V_d > 0$). Accordingly, the effective dopant density in a volume, v, of semiconductor material will be:

$$p = p_0\left(1 - \frac{Q}{qp_0 v}\right) \qquad (9.2)$$

where p_0 is the initial hole density in the organic semiconductor before the application of a gate voltage and Q is the total charge of the cations injected into the conducting polymer film from the electrolyte.

Assuming that the gate electrode is ideally polarizable, the ionic circuit can be described by a resistor (R_s) and a capacitor (C_d) in series.[47] The resistor describes the conductivity of the electrolyte and depends on its ionic strength. The capacitor accounts for polarization at the conducting polymer film/electrolyte and the gate/electrolyte interfaces. Because the capacitance per unit area of a conducting polymer is generally greater than that of the Pt gate,[48] the total capacitance will be determined by the gate capacitance. Because C_d depends on the device area considered, it is convenient to refer to $C_d = c_d \times A$ for much of the analysis, where c_d is the capacitance per unit area and A is the area of the device under consideration.

To solve for drain current, the effective dopant density must be spatially known throughout the organic film. If a differential slice, dx, in the vicinity of position x is considered (Figure 9.4b), then the charge in that slice at steady-state is related to total charge that passes through the circuit:

$$Q(x) = c_d W dx \left(V_g - V(x)\right) \qquad (9.3)$$

where $V(x)$ and W are the spatial voltage profile within the conducting polymer film and the width of the conducting polymer film, respectively. Combining Equations (9.1) through (9.3), it is possible to obtain the governing equation for OECT characteristics at steady state:

$$J(x) = q\mu p_0 \left[1 - \frac{V_g - V(x)}{V_p}\right]\frac{dV(x)}{dx} \qquad (9.4)$$

where V_p is the pinch-off voltage, defined as $q \times p_0 \times T/c_d$.

In the first quadrant of Figure 9.4a ($V_d > 0$), there are two regimes of behavior. First, when $V_d < V_g$, de-doping will occur everywhere in conducting polymer film. Using the previous assumptions, Equation (9.4) can be rewritten in terms of I_{sd} and then solved explicitly, placing the source at $x = 0$ and the drain at $x = L$:

$$I_{sd} = G \left[1 - \frac{V_g - \frac{1}{2} V_d}{V_p} \right] V_d \tag{9.5}$$

where G is the conductance of the conducting polymer film ($G = q \times \mu \times p_0 \times W \times T/L$). The second regime occurs when $V_d > V_g$, and de-doping will only occur in the region of the device where $V(x) < V_g$. This regime is described by:

$$I_{sd} = G \left[V_d - \frac{V_g^2}{2V_p} \right] \tag{9.6}$$

where I_{sd} is linear with V_d, and the onset of linear behavior occurs when $V_d = V_g$.

In the third quadrant ($V_d < 0$), it is possible to completely de-dope portions of the conducting polymer film when the local density of injected cations becomes equal to the intrinsic dopant density of the conducting polymer. This is true when $(V_g - V_d) \geq V_p$, where the critical V_d for saturation can be written as $V_d^{sat} = V_g - V_p$. Locally the semiconductor will be depleted near the drain contact, but holes injected into this region will still be transported to the drain. If the magnitude of V_d increases beyond V_d^{sat}, the extent of the depleted region will move slightly toward the source. In the limit of long channels, for $V_d \leq V_d^{sat}$, I_{sd} will only depend on V_d at saturation for a particular V_g:

$$I_{sd} = -\frac{G(V_d^{sat})^2}{2V_p} \tag{9.7}$$

The model yields an excellent fit (solid lines) to experimental steady-state current–voltage characteristics (points) shown in Figure 9.4a. Such a fit relies on two parameters. The first is the conductance of the conducting polymer film ($G = q \times \mu \times p_0 \times W \times T/L$), which can be determined with conventional techniques such as four-point probe measurement. The second parameter is the pinch-off voltage ($V_p = q \times p_0 \times T/c_d$) and is a measure of the dopant density of the conducting polymer relative to the ionic charge that is leveraged from solution for de-doping. V_p indicates the onset of saturation in the absence of V_g and is akin to the pinch-off voltage in conventional depletion mode field effect transistors.[49,50]

When OECTs are used for sensing applications, it is important to understand the relative, rather than absolute, device response upon gating. Namely, the relevant parameter is the current modulation, $\Delta I_{sd}/I_{sd}$ where ΔI_{sd} is the change in drain current upon application of V_g. As shown in Figure 9.4c, the relative device response is large in the third quadrant of operation and increases with increasing V_g. Such characteristics are paramount in developing high sensitivity sensors and are a useful tool in determining optimal device operating conditions.

9.2.2 OECTs as Electrochemical Sensors

Enzymatic sensing is mostly used for the detection of glucose in human blood.[51] Although inexpensive and portable glucose monitors are commercially available, glucose detection is the fruit fly of enzymatic sensing, and new concepts developed for this purpose can be translated to sensing of other metabolites and disease markers for which there is no commercially available monitor. Moreover, sensors with a lower glucose detection range might allow the non-invasive measurement of glucose.[52]

A simple glucose sensor can be demonstrated using an OECT with PEDOT:PSS channel and source/drain electrodes, and a Pt gate electrode immersed in phosphate buffered saline (PBS), as shown in Figure 9.5a.[21] When both glucose and the enzyme GOx are present in the electrolyte, a large modulation in I_{sd} upon application of V_g is observed. This modulation is much larger than the one obtained when only glucose or only GOx is present in the electrolyte and it is due to the reaction cycle shown in Figure 9.5b. Oxidation of glucose by GOx produces H_2O_2, which can be oxidized to O_2 at the Pt electrode. This reaction is accompanied by de-doping of the PEDOT:PSS channel. To understand how this works, one needs to consider the device operation of OECT in the Faradaic regime.

The addition of glucose and GOx to the OECT affects I_{sd} to an extent that depends on gate voltage and glucose concentration. The transfer characteristics of these

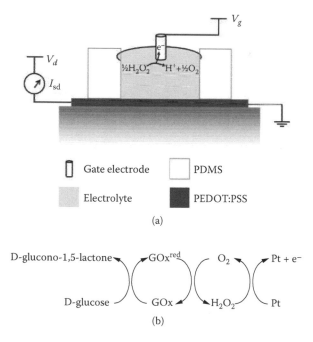

(a)

(b)

FIGURE 9.5 (a) Schematic of a typical organic electrochemical transistor (not drawn to scale). The reaction of interest is shown at the gate electrode. (b) Cycle of reactions involved in glucose sensing. (From Bernards, D.A. et al. Enzymatic sensing with organic electrochemical transistors. *J. Mater. Chem.* **18**, 116–120 (2008). With permission.)

devices (Figure 9.6a) reveal that the magnitude of the modulation increases with increasing glucose concentration. Interestingly, the data in Figure 9.6a can be scaled to yield a universal curve (Figure 9.6b), where the gate voltage is scaled to an effective gate voltage (V_g^{eff}) according to: $V_g^{eff} = V_g + V_{offset}$. V_{offset} is an offset voltage that has a logarithmic dependence on glucose concentration up to 1 mM, which tapers off at higher concentrations. Logarithmic behavior is reminiscent of the Nernst equation that describes the dependence of electrochemical potential on the concentration of redox-active species:[47]

$$E_{Nernst} = E^0 + \frac{kT}{nq} ln \frac{[Ox]}{[Red]} \qquad (9.8)$$

where [Ox] and [Red] are the concentrations of oxidized and reduced species, $E^{0'}$ is the formal potential, k is the Boltzmann constant, T is the absolute temperature, q is the fundamental charge, and n is the number of electrons transferred during the reaction.

The physical meaning of V_{offset} can be understood by comparing OECT sensors with conventional electrochemical sensors. In electrochemistry, the effects of the Nernst equation are manifested by changes of the potential at working electrode where the reaction occurs relative to a fixed potential of reference electrode. In OECTs, the potential shift described by the Nernst equation is manifested by a shift of the electrolyte potential relative to the fixed gate potential. In the non-Faradaic regime, the electrolyte potential (V_{sol}) is determined by the capacitances associated to the double layer formation at the gate (C_{gate}) and channel ($C_{channel}$) and is equal to:

$$V_{sol.} = \frac{V_g}{(1+\gamma)} \qquad (9.9)$$

where γ is defined as $C_{channel}/C_{gate}$.

When both glucose and GOx are added to an electrolyte solution, the reaction shown in Figure 9.5b takes place and the potential drop across the Pt gate/electrolyte interface decreases by the electron transfer from the oxidation of hydrogen peroxide (H_2O_2) at Pt. This Faradaic contribution is described by the Nernst equation:

$$V_{sol.} = \frac{V_g}{(1+\gamma)} + \frac{kT}{2q} ln[H_2O_2] + Constant \qquad (9.10)$$

where the constant contains the details of proton and oxygen activity. This value of the electrolyte potential is described by the dashed line in Figure 9.6c. The oxidation of H_2O_2 at Pt gate increases the electrolyte potential, which in turn decreases the drain current. It is convenient to introduce V_g^{eff} to describe this effect:

$$V_g^{eff} = V_g + (1+\gamma)\frac{kT}{2q} ln[H_2O_2] + Constant \qquad (9.11)$$

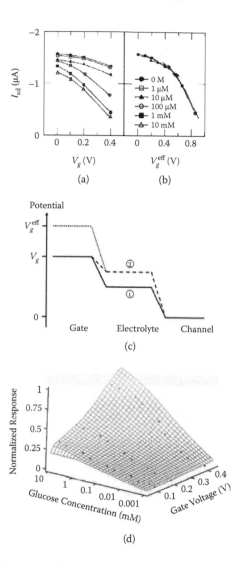

FIGURE 9.6 (a) Drain current plotted as a function of applied gate voltage for a fixed drain voltage ($V_d = -0.2$ V) and various glucose concentrations. (b) Drain current plotted as a function of effective gate voltage, where the applied gate voltage is shifted by a constant that depends on concentration such that the measured current lies along a universal curve. The extent of the shift is determined by glucose concentration. (c) Diagram showing how the potential varies within an enzyme-based OECT. In the absence of reactions, the solution potential *(1)* is determined by the relative capacitances of the gate and the channel. The solution potential in the Faradaic regime *(2)* is increased according to the Nernst equation. The effective gate voltage describes the required gate voltage to achieve the solution potential in the Faradaic regime in the absence of reactions. (d) Plot of relative device response as a function of gate voltage and glucose concentration. Points show experimental device response and surface shows the result of the model. (From Bernards, D.A. et al. Enzymatic sensing with organic electrochemical transistors. *J. Mater. Chem.* **18**, 116–120 (2008). With permission.)

where the new constant is that of Equation (9.10) multiplied by $(1+\gamma)$. V_g^{eff}, illustrated in Figure 9.6c with a dotted line, is the equivalent V_g that needs to be applied in the absence of Faradaic effects to result in the same I_{sd}.

Incorporation of the effective gate voltage in Equations (9.5) and (9.7) allows one to fit the sensor response (Figure 9.6d). The normalized response (NR) of I_{sd} is plotted as a function of glucose concentration and V_g. Normalization was done relative to the zero concentration limitation as:

$$\text{NR} = \left| \frac{I_{sd}^{conc} - I_{sd}^{conc=0}}{I_{sd}^{conc=0}} \right| \tag{9.12}$$

where I_{sd} is considered at zero concentration and at the concentration of interest.

This normalization provides a maximum range of response from zero (no analyte) to one (upper concentration limit) and facilitates comparison between different devices. Figure 9.6d shows experimental data (filled circles) from PEDOT:PSS OECTs along with a fit to Equations (9.5), (9.7), and (9.11), where $\gamma = 4$, $V_p = 0.8$ V, and a correction for the resistivity of the source and drain electrodes is used.

9.2.3   INFLUENCE OF DEVICE GEOMETRY ON SENSOR RESPONSE

The geometry of an OECT plays a role in determining its sensor characteristics in the Faradaic regime. To investigate this role, OECT sensors using planar structure were fabricated in which the ratio between channel and the gate areas ($\gamma = C_{ch}/C_g \sim A_{ch}/A_g$) was varied systematically. The patterning process employed to fabricate the devices is shown in Figure 9.7. After the definition of Pt source, drain, and gate electrodes using a standard photolithographic process, PEDOT:PSS channels were patterned using the parylene lift-off technique.[53] As a final step, the reservoirs for the electrolyte solution were patterned by selective etching of (tridecafluoro-1,1,2,2-tetrahydrooctyl)trichlorosilane (FOTS), whose hydrophobic character confines the electrolyte solution as shown in Figure 9.7d. OECTs with different γ were fabricated on the same substrate (Figure 9.7e).

The current modulation $\Delta I/I_0$ [$\Delta I/I_0 = (|I-I_0|/I_0)$, where I is the off current ($V_g \neq 0$) and the I_0 is the on current ($V_g = 0$)] of OECTs with different γ was evaluated as a function of the H_2O_2 concentration in PBS, and for various values of the V_g. The overall results for two OECTs with $\gamma = 40$ and $\gamma = 0.2$ are shown in Figure 9.8. In the OECT in which the gate is smaller than the channel ($\gamma = 40$, Figure 9.8a), $\Delta I/I_0$ shows a pronounced change with H_2O_2 concentration. The change becomes larger as V_g increases. In the OECT in which the gate is larger than the channel ($\gamma = 0.2$, Figure 9.8b), the change of $\Delta I/I_0$ with H_2O_2 concentration is small, regardless of V_g. These results indicate that a small gate works best for the detection of analytes that interact with the gate electrode.

The influence of the relative size of the gate electrode is emphasized in Figure 9.8c, where the responses of OECTs with different γ are compared at a fixed V_g. Here the plot yields the typical S-shaped curves observed in sensors: For low concentrations ($[H_2O_2] < 10^{-6}$ M), a background $\Delta I/I_0$, which is independent of concentration, is

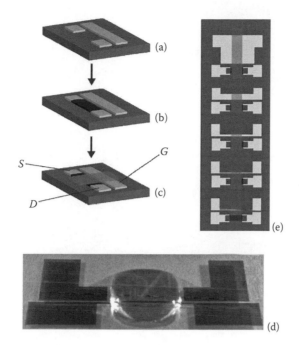

FIGURE 9.7 (a–c) Schematics of the patterning process, (d) image of a device with $\gamma = 40$, and (e) layout of OECTs with $\gamma = 40, 10, 5, 1$, and 0.2 (from bottom to top) on the same substrate. The patterning process involves the definition of Pt source (S), drain (D), and gate (G) electrodes (a), a PEDOT:PSS channel (b), and a hydrophobic SAM that confines the electrolyte over the channel and gate electrode (c).

measured. This regime is followed by a second one (5×10^{-6} M < $[H_2O_2]$ < 10^{-3} M) in which $\Delta I/I_0$ increases with concentration, and the sensitivity (the slope of $\Delta I/I_0$ with concentration) changes with γ. Finally, a third regime is observed ($[H_2O_2] > 10^{-3}$ M), in which $\Delta I/I_0$ saturates.

The data of Figure 9.8c reveal the following information: (1) the background value of $\Delta I/I_0$ decreases with γ, which means that devices with small gates show a small background signal; (2) the sensitivity increases with γ, which means that devices with small gates show high sensitivity; (3) $\Delta I/I_0$ saturates at the same value and for the same concentration independent of channel/gate area; and (4) the detection range does not seem to depend on the channel/gate area ratio.

In order to understand the origin of the background signal, it is important to distinguish between the OECT regimes of operation. In the absence of H_2O_2, there are no charge transfer reactions at the gate electrode for V_g below the threshold of water electrolysis. This is the non-Faradaic regime, in which there is no steady-state gate current.[44] When H_2O_2 is added to the electrolyte reservoir, electrons are transferred to the gate electrode according to the reaction:

$$H_2O_2 \rightarrow O_2 + 2H^+ + 2e^-$$ (9.13)

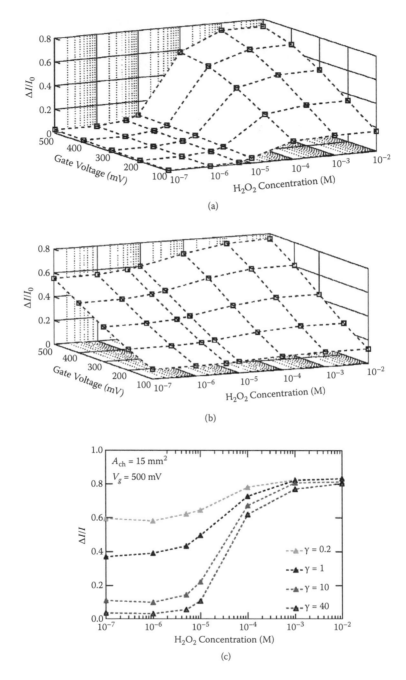

FIGURE 9.8 $\Delta I/I_0$ as a function of H_2O_2 concentration and gate voltage for OECTs with (a) $\gamma = 40$ (small gate) and (b) $\gamma = 0.2$ (large gate), for $V_d = -0.2$ V. (c) Response of OECTs with different values of γ as a function of H_2O_2 concentration for $V_d = -0.2$ V and $V_g = 0.5$ V. The area of the channel is $15\mu m^2$. (d) Non-Faradaic response of OECTs as a function of γ for different gate voltages for $V_d = -0.2$ V.   **(Continued)**

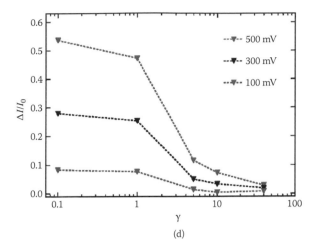

FIGURE 9.8 (Continued)   (d) Non-Faradaic response of OECTs as a function of γ for different gate voltages for $V_d = -0.2$ V.

This is the Faradaic regime. Assuming that the transistor channel is grounded and the gate bias is applied, the electrolyte potential (V_{sol}) can be expressed according to Equation (9.10). The first term on the right-hand side in Equation (9.10) represents the non-Faradaic contribution to V_{sol}, and only depends on V_g and γ; the second represents the Faradaic contribution to V_{sol} as described by the Nernst equation. According to the above, V_{sol} is the potential acting on the OECT channel.

To understand the behavior of OECTs better, we characterized $\Delta I/I_0$ in the non-Faradaic regime. This was achieved in OECTs with just PBS in the electrolyte reservoir, and the results for various geometries and V_g are shown in Figure 9.8d. In accordance with the first term in Equation (9.10), the current modulation decreases strongly with γ. Moreover, the dependence of $\Delta I/I_0$ on V_g also decreases with γ. The significance of these results is the following. In OECTs in which the gate electrode is much smaller than the channel (large γ), most of the applied V_g is dropped at the gate/electrolyte interface. As a result, V_{sol} is small, leading to a small current modulation in the channel. Therefore, the drain current in an OECT with a small gate is only weakly coupled to the applied V_g, and a modulation of the latter will only cause a small modulation of $\Delta I/I_0$. These observations are consistent with the low concentration regions in Figure 9.8a, 9.8b, and 9.8c, and imply that the background signal observed at low H_2O_2 concentrations is of non-Faradaic origin.

A similar argument can be used to understand the dependence of sensitivity on γ. Here the device operates in the Faradaic regime, and the potential drop at the gate/electrolyte interface is reduced by the Nernst term in Equation (9.10). In an OECT with a small gate, where geometry favors a small V_{sol} in the absence of analyte, the modulation caused by the addition of analyte will proportionally larger than in an OECT with a large gate. Therefore, devices with smaller gates will show the highest sensitivity.

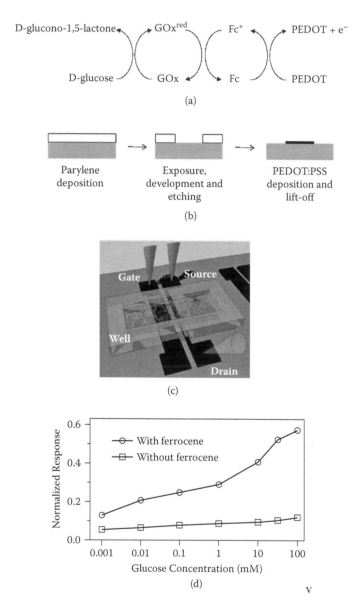

FIGURE 9.9 (a) Reaction cycle for detection of glucose in devices utilizing a PEDOT:PSS electrode and a ferrocene mediator. (b) Diagram of the OECT fabrication process and (c) layout of a finished device. (d) Normalized response to glucose concentration for OECTs pre-loaded with a mixture with (open circles) and without (open squares) ferrocene mediator.

The minimum detectable concentration for OECT-based sensors corresponds to the appearance of the Nernst-type contribution [Equation (9.10)]. The minimum detectable concentration seems to be independent on γ. This can be attributed to the fact that both background signal and sensitivity depend on γ: as the former decreases, the latter increases resulting in a minimum detectable concentration that hovers around 5×10^{-6} M. It is worth noting that the same value for minimum detectable concentration was measured in non-planar OECTs, using a Pt wire as the gate electrode.[21]

The maximum detectable concentration is between 10^{-3} and 10^{-2} M (Figure 9.8c) and also seems to be rather independent of γ. To understand this, one needs to consider the reason for saturation in $\Delta I/I_0$. Saturation happens when the Faradaic contribution to the electrolyte potential becomes comparable to the gate bias. At this limit $V_{sol} \approx V_g$, meaning that the potential applied at the gate electrode drops entirely at the electrolyte/channel interface. Further addition of analyte will not increase the latter and will not result in further de-doping of the PEDOT:PSS channel. Consistent with this conclusion is the fact that the saturation value of $\Delta I/I_0$ increases with gate bias, as seen in Figure 9.8a and 9.8b.

9.2.4   Use of Mediators in the Electrochemical Mode

In the OECTs described previously, the presence of a Pt gate electrode complicates device fabrication and increases the fabrication cost; therefore, it is highly desirable to replace it with a PEDOT:PSS electrode. Electrochemical glucose sensors often replace the O_2/H_2O_2 couple with a fast redox couple, such as the ferrocene [bis (n5-cyclopentandienyl) iron] (Fc)/ferricenium ion couple, in order to overcome issues associated with consumption of oxygen.[54] This redox couple shuttles electrons from the reduced enzyme to the working electrode, creating a new pathway as shown in Figure 9.9a (compare with cycle shown in Figure 9.5b). Given its low redox potential, ferrocene can unload electrons to a PEDOT electrode,[55] creating the opportunity to fabricate OECT-based sensors that consist entirely of conducting polymer. By using this concept, OECT consisting of only PEDOT:PSS without any metal electrodes can be fabricated using a one-layer patterning process and this demonstrates the detection of glucose down to the micromolar range.

Figures 9.9b and 9.9c show a diagram of the OECT fabrication process and the layout of a finished device. PEDOT:PSS was patterned using the parylene lift-off technique. To accommodate the analyte solution, a well made from the poly(dimethlysiloxane) (PDMS) was attached on the glass slide. The well was preloaded with a mixture consisting of PBS, glucose oxidase in PBS (500 units/ml), and 10 mM ferrocene in ethanol. Subsequently, a glucose solution in PBS with concentration from 1μM to 100 mM was added to the well.

Figure 9.9d shows the NR of an OECT with a well preloaded with a mixture that did (open circles) and did not (open squares) contain ferrocene, as a function of glucose concentration. The data were acquired for $V_d = -0.2$ V and $V_g = 0.2$ V and normalization was done relative to the zero-concentration limitation, according to Equation (9.12). When the well is preloaded with a mixture that does not contain ferrocene, the normalized response of OECT (open squares in Figure 9.9d) is small and shows only a small variation across the glucose concentration range.

In contrast, when the well is preloaded with a mixture that contains ferrocene, the normalized response of OECT (open circles in Figure 9.9d) increases dramatically across the glucose concentration range. Namely, NR starts at 0.13 for the 1 μM solution and increases to 0.57 for the 100 mM solution. Adequate change in NR is observed in the 1 to 200 μM range. The results are consistent with the reaction cycle shown in Figure 9.9a, according to which the ferrocene/ferrocenium ion couple mediates electron transfer between the redox enzyme and the PEDOT:PSS gate. In agreement with the model described previously, the flow of electrons to the gate electrode decreases the potential drop at the gate/electrolyte interface. As the gate electrode is held at a fixed bias with respect to the channel, the potential drop at the electrolyte/channel interface increases. The latter results in more effective gating of the transistor channel and the drain current decreases in a way that depends on glucose concentration.

9.3   INTEGRATION OF OECTS WITH MICROFLUIDIC CHANNELS

The term microfluidics refers to devices and systems for manipulating the small amounts of fluid (10^{-9} to 10^{-18} l) using channels with characteristic dimensions of tens to hundreds of micrometers.[56] Fluidic systems at the micrometer scale display low Reynolds numbers and the flow is governed mainly by viscous forces and pressure gradients with low moments of inertia. This indicates that fluid flowing in a micron-scale channel takes place in a turbulence-free fashion.[57] One consequence of laminar flow in microchannels is that two or more streams that come together flow parallel to each other without turbulence, and the diffusion of molecules across the interface between fluids is the only way for mixing.[57,58] Also, high surface-to-volume ratios, small diffusion distances, and surface tension become dominant in microfluidics due to the small length scale.[58] There have been rapid developments of microfluidic components such as pneumatically activated valves, mixers, pumps, and sensors for methods of introducing, moving, mixing, and detecting reagents and samples.[59] The full integration of those components can help realize the concept of the lab-on-a-chip, in which chemical synthesis, reactions, analysis, and sensing are carried out on a single chip using only small fluid volumes. Because the unique physical behavior of microfluidic systems allows for functionalities that are almost impossible to access on a macroscopic scale, microfluidic systems can provide specific advantages such as low fluid volume consumption, faster response times due to the small diffusion distances and high surface-volume ratios, compactness due to the integration of many functionalities in small volume, and lower fabrication costs.[56,58–60]

Mabeck et al. reported a microfluidic gating technique that demonstrates how OECTs could be used as transducers in microfluidic sensors.[61] In this work, the gate electrode of the OECT was integrated into the ceiling of PDMS microchannel. A modulation of the drain current along a PEDOT:PSS channel was demonstrated (Figure 9.10). The gold gate electrode used in this work can offer a convenient platform for immobilizing a wide array of specific recognition elements such as DNA, antibodies, and viruses via thiol chemistry.

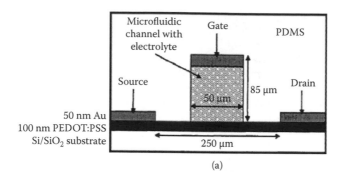

(a)

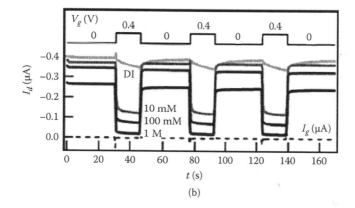

(b)

FIGURE 9.10 (a) Schematic cross-section of the device structure (not to scale) showing the gate electrode integrated into the microfluidic channel. (b) Drain current as a function of time with $V_d = -0.1$ V for DI water and NaCl solutions with varying concentrations. The dotted line is the gate current in 1 M NaCl. (From Mabeck, J.T. et al. Microfluidic gating of an organic electrochemical transistor. *Appl. Phys. Lett.* **87**, 3 (2005). With permission.)

9.3.1 INTEGRATION OF OECT ARRAYS WITH SURFACE-DIRECTED MICROFLUIDIC SYSTEM FOR MULTI-ANALYTE SENSING

One of the novel microfluidic techniques for the manipulation of fluid flow is to control the wettability of a surface by patterning its surface energy.[62–65] Beebe and coworkers have developed the surface-directed flow by patterning hydrophilic and hydrophobic regions inside microchannel networks with the self-assembled monolayer (SAM) in combination with either multi-stream laminar flow or photolithography.[64,65] Such surface-patterned microfluidic channels consist of a hydrophilic microfluidic pathway and a hydrophobic "virtual" wall. When a drop of aqueous solution is placed on such a patterned surface, it is confined by the hydrophobic virtual wall and allowed to flow only along the hydrophilic pathway. A unique feature of this concept is that the capillary action generated by the hydrophilic surface provides the driving force for spontaneous flow in the microfluidic channel. Therefore, no external pressure such as a syringe pump is required to direct the flow.[65]

To achieve a multianalyte sensing system, the integration of OECT arrays with the surface-directed microfluidic system that accepts a single analyte solution and distributes it to several OECT sensors is required.[66] Since most analyte solutions of interest to biological sensing are aqueous ones, a surface-directed microfluidic system with hydrophilic pathways and hydrophobic virtual walls needs to be employed. The surface-directed microfluidic channel in this case was defined by subtractive patterning of FOTS monolayer combined with oxygen plasma etching. Etched region by oxygen plasma formed the hydrophilic pathway and the remaining hydrophobic FOTS layer (contact angle of water ~100°) plays a role for virtual wall to confine aqueous liquid. Figure 9.11a and 9.11b show how to fabricate the surface-directed microfluidic channel and the microfluidic system in action by spontaneous flow of aqueous solution along a 500-μm wide hydrophilic microchannel. The microfluidic system consists of one input reservoir and four measurement reservoirs where OECT sensors are placed. The resulting OECT component shown in Figure 9.11c consists of Pt source and drain electrodes and a 300-μm wide PEDOT:PSS channel. A common Pt gate electrode runs along all four measurement reservoirs—it is 300 μm wide and 400 μm away from the OECT active channels. The hydrophilic measurement reservoir is an integral part of the OECT as it confines the analyte solution over the active area including PEDOT:PSS channel and gate electrode. It is also worth noting that sample volumes as small as 5 μl are adequate to fill the microfluidic system.

The integration of the surface-directed microfluidic system with the OECT array requires careful optimization of the fabrication process because one needs to retain both the spontaneous flow in the microfluidic system and avoid degradation of the OECT array performance. The best way to achieve this is to fabricate the hydrophilic channel as the last step in order to avoid contamination that prevents spontaneous flow, as described in Figure 9.12a. Fabrication of OECT arrays started with patterning source/drain and gate Pt electrodes on glass substrate, followed by FOTS deposition and PEDOT:PSS patterning with parylene lift-off technique. To avoid damage of patterned PEDOT:PSS active layer during the O_2 plasma etching for the hydrophilic channels and reservoirs, the mask for the subtractive pattering of the FOTS was designed to leave the PEDOT:PSS channel covered with photoresist. This photoresist protected the PEDOT:PSS channel from the etching process and was removed during the last fabrication step without noticeable damage to the PEDOT:PSS channel. The response of the OECTs was measured after fabrication by applying a drop of PBS in the input reservoir. The solution was guided toward the measurement reservoirs and the flow stopped at the PEDOT:PSS channels. The PEDOT:PSS channels were not hydrophilic enough to allow spontaneous flow, and the lower part of the input reservoir remained dry (including the PEDOT:PSS channel). Addition of a small amount (approximately 1 μl) of PBS on top of each PEDOT:PSS channel completely filled the measurement reservoirs and covered the PEDOT:PSS channels. Figure 9.12b shows a typical response of an OECT to a train of gate pulses. The drain current is shown to decrease every time V_g is applied, in agreement with the model of operation described previously. The magnitude of the current modulation reaches 20% for V_g of 0.5 V, a strong modulation that indicates that the OECTs withstood the fabrication process.

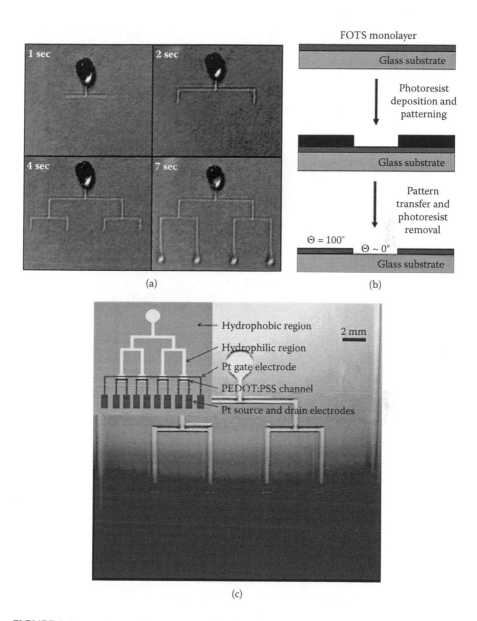

FIGURE 9.11 (a) The fluidic system in action after a drop of water was added to the input reservoir at $t = 0$ sec. (b) Fabrication process for the surface-directed microfluidic channel. (c) Photograph of the fabricated chip with a drop of water distributed along the surface-directed microfluidic system. The inset explains the schematic structure of the chip. (From Yang, S.Y. et al. Integration of a surface-directed microfluidic system with an organic electrochemical transistor array for multi-analyte biosensors. *Lab Chip* **9**, 704–708 (2009). With permission.)

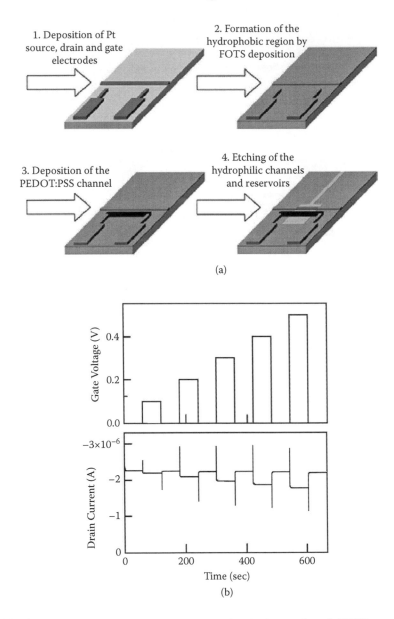

FIGURE 9.12 (a) Optimized fabrication process for the integration of OECT arrays with surface-directed microfluidic system. (b) Modulation of the drain current of an OECT upon the application of a gate voltage. (From Yang, S.Y. et al. Integration of a surface-directed microfluidic system with an organic electrochemical transistor array for multi-analyte biosensors. *Lab Chip* **9**, 704–708 (2009). With permission.)

(Continued)

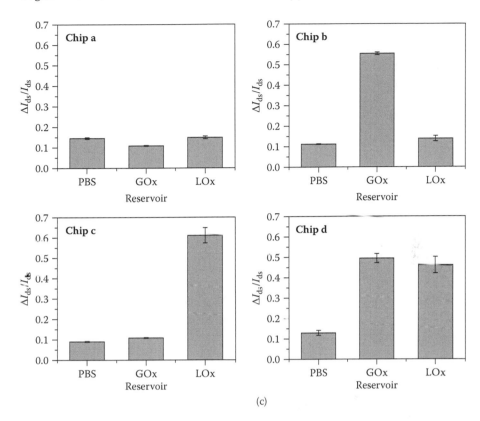

(c)

FIGURE 9.12 (Continued)   (c) Relative modulation of the drain current in four chips, each of which was exposed to a different solution. V_d and V_g were −0.2V and 0.3V, respectively. Each chip contained OECTs with PBS, GOx, or LOx added in their measurement reservoirs. The error bars refer to the repeated measurements on the same OECT. (From Yang, S.Y. et al. Integration of a surface-directed microfluidic system with an organic electrochemical transistor array for multi-analyte biosensors. *Lab Chip* **9**, 704–708 (2009). With permission.)

Figure 9.12c shows the multianalyte sensing capability of the integrated surface-directed microfluidic/OECT array. Four identical chips were fabricated and exposed to different solutions, including PBS alone (chip **a**), a 10-mM glucose solution in PBS (chip **b**), a 10-mM lactate solution in PBS (chip **c**), and a PBS solution that contained 10 mM of glucose and 10 mM of lactate (chip **d**). For each chip, a solution was added to the input reservoir and was allowed to reach the measurement reservoirs. Subsequently, 1 µl of solutions of GOx (250 units/ml) and lactate oxidase (LOx, 250 units/ml) in PBS were added on top of the PEDOT:PSS stripes of the second and third measurement reservoirs, respectively. An equal amount of PBS alone was added on top of the PEDOT:PSS stripe of the first reservoir, which served as the reference. As shown in Figure 9.12c, all OECTs show a rather small modulation upon exposure to pure PBS (chip **a**).

A similarly small modulation is registered in OECTs that are exposed to analyte solutions, but do not contain the appropriate enzyme in their reservoirs. Exposure of a transistor to both the analyte and the appropriate enzyme such as glucose and GOx or lactate and LOx generates a considerably larger response (chip **b** and **c** in Figure 9.12c), which allows the identification of an analyte in an unknown solution. In addition, the simultaneous detection of glucose and lactate was demonstrated (chip **d** in Figure 9.12c), illustrating the potential of this concept for the development of multi-analyte sensor components for lab-on-a-chip applications. It is also worth noting that enzymes can be entrapped into the conducting polymer—the parylene lift-off technique can be repeated several times to pattern a different enzyme trapped-conducting polymer in each measurement reservoir.[67]

9.4   CONCLUDING REMARKS

OECT-based sensors have been fabricated with different materials and have been used to detect a wide range of biological and chemical species. Modeling has provided an improved understanding that paves the way for rational device optimization. Ease of processing and low cost, together with tunability of electronic properties and integration with biological systems make OECTs ideal candidates for biosensing applications. The response of H_2O_2 sensors based on PEDOT:PSS OECTs can be tuned by varying the ratio between the areas of the channel and the gate electrode. Devices with small gate electrodes were shown to have a lower background signal and higher sensitivity. These results point to important lessons on how to optimize materials and device geometry in organic electrochemical transistors for sensor applications. Also, a glucose sensor based on PEDOT:PSS OECTs that are made only of PEDOT:PSS without any metal electrodes was demonstrated. The device employed a ferrocene mediator to shuttle electrons between the enzyme GOx and the PEDOT:PSS gate electrode and offers a simple architecture for enzymatic sensing that can be fabricated using a one-layer patterning process. The integration of OECT arrays with surface-directed microfluidics can help realize a single chip in which the multi-analyte solution can be distributed to a number of OECT sensors without any application of external pressure. The resulting simultaneous detection of many metabolites on one chip illustrates the potential of this integration concept for the development of muti-analyte sensor components for lab-on-a-chip application.

The OECTs sensors described here are able to detect micromolar glucose or H_2O_2 concentrations. Such values are consistent with clinical glucose levels found in human saliva.[52] This detection limit can be improved further by adapting techniques developed in electrochemistry, which have led to detection of very low H_2O_2 concentrations. In cyclic voltammetry, for instance, use of a chemically activated redox mediator[68] or electrodes modified with nanoparticles[69] have allowed detection in the picomolar and nanomolar range, respectively. Such approaches can be readily adapted to OECT design to lower the minimum detectable analyte concentration.

Finally, our understanding of OECT-based sensors will benefit from a better understanding of the distribution of ions at the interface between conducting polymers and electrolytes.[70] Of particular interest is the ratio between ions that accumulate at the surface vs. ions that enter in the polymer film, as this ratio ultimately determines the "ease" with which a conducting polymer can be de-doped (hence, it controls the maximum detectable analyte concentration). Most conducting polymers have been optimized for electronic conductivity, and improvements can be achieved by the simultaneous optimization of permeability to ions.

REFERENCES

1. Dimitrakopoulos, C.D. & Malenfant, P.R.L. Organic thin film transistors for large area electronics. *Adv. Mater.* **14**, 99 (2002).
2. Dodabalapur, A. Organic and polymer transistors for electronics. *Mater. Today* **9**, 24–30 (2006).
3. Braga, D. & Horowitz, G. High-performance organic field-effect transistors. *Adv. Mater.* **21**, 1473–1486 (2009).
4. Volkel, A.R., Street, R.A. & Knipp, D. Carrier transport and density of state distributions in pentacene transistors. *Phys. Rev. B* **66**, 8 (2002).
5. Panzer, M.J. & Frisbie, C.D. High carrier density and metallic conductivity in poly(3-hexylthiophene) achieved by electrostatic charge injection. *Adv. Funct. Mater.* **16**, 1051–1056 (2006).
6. Said, E. et al. Polymer field-effect transistor gated via a poly(styrenesulfonic acid) thin film. *Appl. Phys. Lett.* **89**, 3 (2006).
7. Herlogsson, L. et al. Low-voltage polymer field-effect transistors gated via a proton conductor. *Adv. Mater.* **19**, 97 (2007).
8. Herlogsson, L. et al. Downscaling of organic field-effect transistors with a polyelectrolyte gate insulator. *Adv. Mater.* **20**, 4708 (2008).
9. Panzer, M.J. & Frisbie, C.D. Exploiting ionic coupling in electronic devices: Electrolyte-gated organic field-effect transistors. *Adv. Mater.* **20**, 3177–3180 (2008).
10. Mitra, S., Shukla, A.K. & Sampath, S. Electrochemical capacitors with plasticized gel-polymer electrolytes. *J. Power Sources* **101**, 213–218 (2001).
11. Panzer, M.J. & Frisbie, C.D. Polymer electrolyte-gated organic field-effect transistors: Low-voltage, high-current switches for organic electronics and testbeds for probing electrical transport at high charge carrier density. *J. Am. Chem. Soc.* **129**, 6599–6607 (2007).
12. Yuen, J.D. et al. Electrochemical doping in electrolyte-gated polymer transistors. *J. Am. Chem. Soc.* **129**, 14367–14371 (2007).
13. Dhoot, A.S. et al. Beyond the metal-insulator transition in polymer electrolyte gated polymer field-effect transistors. *Proc. Natl. Acad. Sci. U.S.A.* **103**, 11834–11837 (2006).
14. Nilsson, D. et al. Bi-stable and dynamic current modulation in electrochemical organic transistors. *Adv. Mater.* **14**, 51–54 (2002).
15. Zhu, Z.T. et al. A simple poly(3,4-ethylene dioxythiophene)/poly(styrene sulfonic acid) transistor for glucose sensing at neutral pH. *Chem. Commun.*, 1556–1557 (2004).
16. Nilsson, D., Robinson, N., Berggren, M. & Forchheimer, R. Electrochemical logic circuits. *Adv. Mater.* **17**, 353 (2005).
17. Mabeck, J.T. & Malliaras, G.G. Chemical and biological sensors based on organic thin-film transistors. *Anal. Bioanal. Chem.* **384**, 343–353 (2006).
18. Bernards, D.A., Malliaras, G.G., Toombes, G.E.S. & Gruner, S.M. Gating of an organic transistor through a bilayer lipid membrane with ion channels. *Appl. Phys. Lett.* **89**, 3 (2006).
19. Berggren, M. & Richter-Dahlfors, A. Organic bioelectronics. *Adv. Mater.* **19**, 3201–3213 (2007).
20. Macaya, D.J. et al. Simple glucose sensors with micromolar sensitivity based on organic electrochemical transistors. *Sens. Actuator B-Chem.* **123**, 374–378 (2007).
21. Bernards, D.A. et al. Enzymatic sensing with organic electrochemical transistors. *J. Mater. Chem.* **18**, 116–120 (2008).

22. White, H.S., Kittlesen, G.P. & Wrighton, M.S. Chemical derivatization of an array of 3 gold microelectrodes with polypyrrole - fabrication of a molecule-based transistor. *J. Am. Chem. Soc.* **106**, 5375–5377 (1984).

23. Bartlett, P.N. & Birkin, P.R. A microelectrochemical enzyme transistor responsive to glucose. *Anal. Chem.* **66**, 1552–1559 (1994).

24. Bartlett, P.N., Birkin, P.R., Wang, J.H., Palmisano, F. & De Benedetto, G. An enzyme switch employing direct electrochemical communication between horseradish peroxidase and a poly(aniline) film. *Anal. Chem.* **70**, 3685–3694 (1998).

25. Bartlett, P.N. & Wang, J.H. Electroactivity, stability and application in an enzyme switch at pH 7 of poly(aniline)-poly(styrenesulfonate) composite films. *J. Chem. Soc.-Faraday Trans.* **92**, 4137–4143 (1996).

26. Bartlett, P.N., Wang, J.H. & James, W. Measurement of low glucose concentrations using a microelectrochemical enzyme transistor. *Analyst* **123**, 387–392 (1998).

27. Chao, S.H. & Wrighton, M.S. Characterization of a solid-state polyaniline-based transistor - water-vapor dependent characteristics of a device employing a polyvinyl-alcohol) phosphoric-acid solid-state electrolyte. *J. Am. Chem. Soc.* **109**, 6627–6631 (1987).

28. Dabke, R.B., Singh, G.D., Dhanabalan, A., Lal, R. & Contractor, A.Q. An ion-activated molecular electronic device. *Anal. Chem.* **69**, 724–727 (1997).

29. Paul, E.W., Ricco, A.J. & Wrighton, M.S. Resistance of polyaniline films as a function of electrochemical potential and the fabrication of polyaniline-based microelectronic devices. *J. Phys. Chem.* **89**, 1441–1447 (1985).

30. Saxena, V., Shirodkar, V. & Prakash, R. Copper(II) ion-selective microelectrochemical transistor. *J. Solid State Electrochem.* **4**, 234–236 (2000).

31. Kanungo, M., Srivastava, D.N., Kumar, A. & Contractor, A.Q. Conductimetric immunosensor based on poly(3,4-ethylenedioxythiophene). *Chem. Commun.*, 680–681 (2002).

32. Krishnamoorthy, K., Gokhale, R.S., Contractor, A.Q. & Kumar, A. Novel label-free DNA sensors based on poly(3,4-ethylenedioxythiophene). *Chem. Commun.*, 820–821 (2004).

33. Belanger, D., Nadreau, J. & Fortier, G. Electrochemistry of the polypyrrole glucose-oxidase electrode. *J. Electroanal. Chem.* **274**, 143–155 (1989).

34. Bartlett, P.N., Wang, J.H. & Wallace, E.N.K. A microelectrochemical switch responsive to NADH. *Chem. Commun.*, 359–360 (1996).

35. Ballarin, B. et al. Thermal inkjet microdeposition of PEDOT:PSS on ITO-coated glass and characterization of the obtained film. *Synth. Met.* **146**, 201–205 (2004).

36. Mabrook, M.F., Pearson, C. & Petty, M.C. Inkjet-printed polymer films for the detection of organic vapors. *IEEE Sens. J.* **6**, 1435–1444 (2006).

37. Percin, G., Lundgren, T.S. & Khuri-Yakub, B.T. Controlled ink-jet printing and deposition of organic polymers and solid particles. *Appl. Phys. Lett.* **73**, 2375–2377 (1998).

38. Setti, L. et al. An amperometric glucose biosensor prototype fabricated by thermal ink-jet printing. *Biosens. Bioelectron.* **20**, 2019–2026 (2005).

39. Ashizawa, S., Horikawa, R. & Okuzaki, H. Effects of solvent on carrier transport in poly(3,4-ethylenedioxythiophene)/poly(4-styrenesulfonate). *Synth. Met.* **153**, 5–8 (2005).

40. Crispin, X. et al. The origin of the high conductivity of poly(3,4-ethylenedioxythiophene)-poly(styrenesulfonate) (PEDOT- PSS) plastic electrodes. *Chem. Mat.* **18**, 4354–4360 (2006).

41. Kim, J.Y., Jung, J.H., Lee, D.E. & Joo, J. Enhancement of electrical conductivity of poly(3,4-ethylenedioxythiophene)/poly(4-styrenesulfonate) by a change of solvents. *Synth. Met.* **126**, 311–316 (2002).

42. Ouyang, J. et al. On the mechanism of conductivity enhancement in poly (3,4-ethylenedioxythiophene): poly(styrene sulfonate) film through solvent treatment. *Polymer* **45**, 8443–8450 (2004).

43. A. M. D. Wan, D. J. Brooks, A. Gumus, C. Fischbach & Malliaras, G.G. Electrical control of cell density gradients on a conducting polymer surface. *Chem. Commun.* 5278–5280 (2009).
44. Bernards, D.A. & Malliaras, G.G. Steady-state and transient behavior of organic electrochemical transistors. *Adv. Funct. Mater.* **17**, 3538–3544 (2007).
45. Isaksson, J. et al. Electronic control of Ca2+ signalling in neuronal cells using an organic electronic ion pump. *Nat. Mater.* **6**, 673–679 (2007).
46. M. Berggren et al. *Organic Semiconductors in Sensor Applications*, Vol. 107. (Springer, Berlin; 2008).
47. Bard, A.J. & Faulkner, L.R. *Electrochemical Methods.* (John Wiley and Sons, New York; 1981).
48. Stenger-Smith, J.D. et al. Poly(3,4-alkylenedioxythiophene)-based supercapacitors using ionic liquids as supporting electrolytes. *J. Electrochem. Soc.* **149**, A973–A977 (2002).
49. Middlebrook, R.D. A simple derivation field-effect transistor characteristics. *Proc. IEEE* **51**, 1146 (1963).
50. Wallmark, J.T. The field-effect transistor—a review. *Rca Review* **24**, 641–660 (1963).
51. Heller, A. Implanted electrochemical glucose sensors for the management of diabetes. *Annu. Rev. Biomed. Eng.* **1**, 153–175 (1999).
52. Yamaguchi, M., Mitsumori, M. & Kano, Y. Noninvasively measuring blood glucose using saliva. *IEEE Eng. Med. Biol. Mag.* **17**, 59–63 (1998).
53. DeFranco, J.A., Schmidt, B.S., Lipson, M. & Malliaras, G.G. Photolithographic patterning of organic electronic materials. *Org. Electron.* **7**, 22–28 (2006).
54. Cass, A.E.G. et al. Ferrocene-mediated enzyme electrode for amperometric determination of glucose. *Anal. Chem.* **56**, 667–671 (1984).
55. Doherty, W.J., Armstrong, N.R. & Saavedra, S.S. Conducting polymer growth in porous sol-gel thin films: Formation of nanoelectrode arrays and mediated electron transfer to sequestered macromolecules. *Chem. Mat.* **17**, 3652–3660 (2005).
56. Whitesides, G.M. The origins and the future of microfluidics. *Nature* **442**, 368–373 (2006).
57. Pihl, J., Sinclair, J., Karlsson, M. & Orwar, O. Microfluidics for cell-based assays. *Mater. Today* **8**, 46–51 (2005).
58. Beebe, D.J., Mensing, G.A. & Walker, G.M. Physics and applications of microfluidics in biology. *Annu. Rev. Biomed. Eng.* **4**, 261–286 (2002).
59. Stone, H.A., Stroock, A.D. & Ajdari, A. Engineering flows in small devices: Microfluidics toward a lab-on-a-chip. *Annu. Rev. Fluid Mech.* **36**, 381–411 (2004).
60. El-Ali, J., Sorger, P.K. & Jensen, K.F. Cells on chips. *Nature* **442**, 403–411 (2006).
61. Mabeck, J.T. et al. Microfluidic gating of an organic electrochemical transistor. *Appl. Phys. Lett.* **87**, 3 (2005).
62. Bouaidat, S. et al. Surface-directed capillary system; theory, experiments and applications. *Lab Chip* **5**, 827–836 (2005).
63. Lam, P., Wynne, K.J. & Wnek, G.E. Surface-tension-confined microfluidics. *Langmuir* **18**, 948–951 (2002).
64. Zhao, B., Moore, J.S. & Beebe, D.J. Surface-directed liquid flow inside microchannels. *Science* **291**, 1023–1026 (2001).
65. Zhao, B., Moore, J.S. & Beebe, D.J. Principles of surface-directed liquid flow in microfluidic channels. *Anal. Chem.* **74**, 4259–4268 (2002).
66. Yang, S.Y. et al. Integration of a surface-directed microfluidic system with an organic electrochemical transistor array for multi-analyte biosensors. *Lab Chip* **9**, 704–708 (2009).
67. Gregg, B.A. & Heller, A. Cross-linked redox gels containing glucose-oxidase for amperometric biosensor applications. *Anal. Chem.* **62**, 258–263 (1990).
68. Lyon, J.L. & Stevenson, K.J. Picomolar peroxide detection using a chemically activated redox mediator and square wave voltammetry. *Anal. Chem.* **78**, 8518–8525 (2006).

69. Salimi, A., Hallaj, R., Soltanian, S. & Mamkhezri, H. Nanomolar detection of hydrogen peroxide on glassy carbon electrode modified with electrodeposited cobalt oxide nanoparticles. *Anal. Chim. Acta* **594**, 24–31 (2007).

70. Wang, J. & Bard, A.J. On the absence of a diffuse double layer at electronically conductive polymer film electrodes. Direct evidence by atomic force microscopy of complete charge compensation. *J. Am. Chem. Soc.* **123**, 498–499 (2001).

10 Polyelectrolyte-Gated Organic Field-Effect Transistors

Xavier Crispin, Lars Herlogsson, Oscar Larsson, Elias Said, and Magnus Berggren

CONTENTS

10.1 INTRODUCTION

Organic field-effect transistors (OFETs)[1,2] and other "plastic" electronic devices[3] are currently scrutinized for use in printed, flexible, integrated electronics and displays. Ideally, these systems are fast, operate at low voltage, and are robust enough to be manufactured using standard printing techniques.[1,4,5] Current printing technology allows for a separation between the source and drain electrodes in OFETs of less than 1 μm.[6] In traditional OFETs, the organic semiconductor film is separated from the gate electrode by a thin insulating dielectric film. The gate-insulator-semiconductor sandwich can be seen as a capacitor, where the charge density in the semiconductor, and thus the conductance of the semiconductor channel, is tuned by the applied voltage. Tremendous efforts have been devoted to reach high capacitance (per area) C_i between the gate and the channel to allow transistors to operate at low voltage.[7,8] Since the dielectric constant k of organic materials usually is quite low, very thin gate insulator layers are required in order to obtain a high capacitance. Molecular assembly and self-organization techniques have been utilized to manufacture gate dielectric layers only a few nanometers thick, resulting in large capacitance (C_i up to 1 μF cm^{-2}).[8–10]

Using an electrolyte, for example, salt in a polymer matrix, to gate a silicon-based transistor[11] or a polymer-based electrochemical transistor[12,13] was first demonstrated more than two decades ago. Recently, polymer-based transistors have been gated with a polysaccharide derivative containing LiBF$_4$,[14] or with LiClO$_4$ in polyethylene oxide[15–19] and poly(3,4-ethylenedioxythiophene)-poly(styrenesulfonic acid).[20,21] In these devices, the redistribution of mobile ions within the conjugated polymer layer and the electrolyte together with the charge injection from the source/drain electrodes leads to electrochemical doping (undoping) of the polymer channel. Bulk electrochemistry controls the on- and off-state of the transistor. We classify such transistors as electrochemical transistors. They operate at low voltage (< 2 V) but switch slowly (requiring up to several seconds).

Recently, electrolytes have been explored as gate "insulators" for OFETs. Various electrolyte systems, including hygroscopic insulators (e.g., polyvinylphenol),[22,23] polymer electrolytes,[24] polyelectrolytes,[25,26] ionic liquids,[27] and ion gels,[28,29] have recently been demonstrated in low-voltage OFETs. The transistor works as a field-effect transistor rather than an electrochemical transistor if the bulk electrochemical doping is prevented. This is typically realized by using large molecular or polymeric ions that cannot penetrate into the semiconductor channel. In this chapter, we focus on polyelectrolytes as insulators to gate p-channel OFETs. In Section 10.2, the polyelectrolyte-based capacitors are presented to help identify the various polarization mechanisms involved when an electrical bias is applied to the gate in the OFETs. After describing typical electrical characteristics of polyelectrolyte-gated OFETs in Section 10.3, we investigate various key issues governing their performances in Section 10.4: the role of the size of the anions, the limitation mechanisms on the transistor speed, and the polarization mechanisms in the polyelectrolyte. In Section 10.5, two unique features of polyelectrolyte-gated OFETs are pinpointed: the independence of the polyelectrolyte thickness on the operating voltage and the large electric field created at the polyelectrolyte-semiconductor interfaces. Those specific properties solve two technological issues that are not easily solved with common dielectrics: compatibility to printing technologies and short-channel effects. Finally, ring oscillators serve as benchmarks to test the potential of polyelectrolyte-gated OFETs in electronic circuits.

10.2   POLYELECTROLYTE-BASED CAPACITORS

Beside their necessity for electrochemical reactions, an interesting feature of electrolytes is their ability to form electric double layer capacitors (EDLCs). When a difference in electric potential is applied between two ion-blocking electrodes sandwiching a common electrolyte, the anions (cations) in the electrolyte move toward the positively (negatively) charged electrode to form an electric double layer (EDL), composed of a compact (Helmholtz) layer and a diffuse layer. Such EDLCs can have extraordinarily high capacitance (e.g., C_i = 500 μF cm^{-2})[30] and respond quickly; the result of a charge-separation of only a few angstroms within the Helmholtz layers formed in a few tens of microseconds.[31]

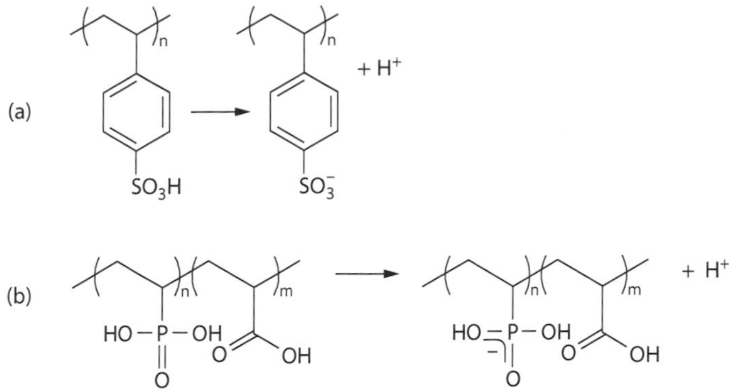

FIGURE 10.1 Chemical structures and dissociation reactions of (a) PSSA and (b) P(VPA-AA).

The polyelectrolytes used in this work are (1) the homopolymer polystyrene sulfonic acid (PSSA) (Figure 10.1a), and (2) a random copolymer of vinylphosphonic acid and acrylic acid, P(VPA-AA) (Figure 10.1b). The sulfonic acid and phosphonic acid groups are strongly acidic (acidic constant pK_a about 2 in water), thus providing plenty of potentially mobile protons. The two polyelectrolytes possess similar properties in capacitor with a capacitance of about 20 µFcm^{-2} at 100 Hz. Capacitors made of a thin PSSA layer (80 nm) sandwiched between two titanium electrodes were fabricated and characterized with impedance spectroscopy. In order to facilitate the correlation between the properties of the polyelectrolyte capacitors and those of the polyelectrolyte-gated transistors, the total complex impedance of the capacitors ($Z_{Tot}(f)$) is described in terms of the effective capacitance per area ($C_{Eff}(f)$) and the phase angle ($\theta(f)$) as functions of the frequency (f). In Figure 10.2, the effective capacitance per area and the phase angle versus the frequency are given for a PSSA capacitor at 10% relative humidity (RH). Based on the value of the phase angle, the relaxation phenomena can be classified with either a capacitive or a resistive character. Three different frequency regions are identified:[32] (1) The capacitive behavior ($\theta(f) < -45°$) at high frequencies ($f > 120$ kHz) is attributed to dipolar relaxation of the material. (2) The resistive behavior ($\theta(f) > -45°$) at intermediate frequencies (800 Hz $< f < 120$ kHz) originates from dissociated protons migrating away from the polymer chains in the oscillating electric field. This is called the ionic relaxation. (3) At lower frequencies ($f < 800$ Hz), the impedance acquires a capacitive character ($\theta(f) < -45°$). Upon decreasing the frequency, the protons migrate long enough to reach the vicinity of the metal electrodes; thus creating the Helmholtz double-layers at the two polyelectrolyte/ metal electrode interfaces. The transition between the resistive (2) and capacitive (3) regimes is accompanied with a sudden increase of the capacitance, which reaches a large value ($C_{Eff}(f)$ ~20 µF cm^{-2} at 100 Hz), characteristic of EDLCs.[30]

10.3　POLYELECTROLYTE-GATED FIELD-EFFECT TRANSISTORS

The idea is to employ the large capacitance of the electrical double layer at the interface between the positively charged semiconductor channel and the polyelectrolyte

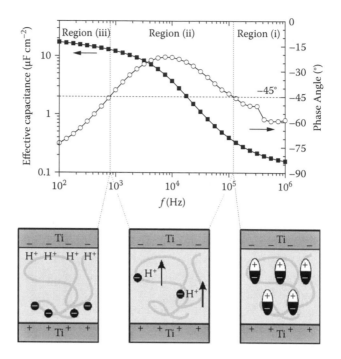

FIGURE 10.2   The effective capacitance per area and the phase angle vs. frequency at 10% RH for a polyelectrolyte capacitor (an 80-nm thick PSSA layer sandwiched between two titanium electrodes). The graph is divided into the three frequency regions of the polyelectrolyte: (1) dipolar relaxation, (2) ionic relaxation, and (3) electric double-layer formation. The origin of these three relaxation mechanisms is illustrated below the graph.

in an OFET. Polyelectrolyte-gated organic field-effect transistors can thus also be called OFETs gated via an electric double layer capacitor (EDLC-OFETs). A polythiophene derivative (P(T$_0$T$_0$TT$_{16}$), see Figure 10.3a), with a relatively large ionization potential (5.1 eV), is used as semiconductor,[33] and the polyanionic electrolyte P(VPA-AA) is chosen as the gate insulator. The electrical characteristics of the OFETs are shown in Figures 10.3c and 10.3d. The output characteristics of the EDLC-OFETs (Figure 10.3c) show clear current modulation for drive voltages of less than 1.2 V. The large saturation current obtained for low operation voltage (V_G = –1 V) reflects the large capacitance of P(VPA-AA). The relationship between those quantities is interpreted via conventional semiconductor theory that reproduces the general trends in OFET characteristics:[34,35]

$$I_D^{sat} = \frac{W}{2L} \mu\, C_i \left(V_G - V_T\right)^2 \tag{10.1}$$

where C_i is the capacitance per unit area of the gate insulator, μ is the charge carrier mobility, and V_T is the threshold voltage. The P(VPA-AA) gate insulator (70 nm thick) has, when sandwiched between the Ti gate electrode and the semiconductor,

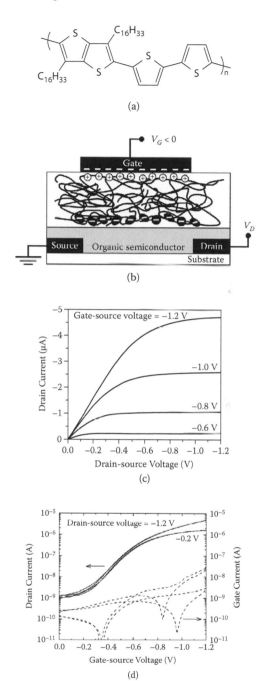

FIGURE 10.3 (a) Chemical structure of $P(T_0T_0TT_{16})$. (b) Schematic drawing of an EDLC-OFET with P(VPA-AA) gate insulator when applying a negative gate voltage, illustrating the proposed model. (c) Output and (d) transfer characteristics. The transistor has a channel length of 2.5 μm and a channel width of 1 mm. Both forward and reverse sweeps are shown.

an effective capacitance of approximately 3 μF cm^{-2}. The carrier mobility can then be calculated to be ~0.02 cm^2 V^{-2} s^{-1}, which is one order of magnitude lower than in prior reports for $P(T_0T_0TT_{16})$-based OTFTs likely due to a low crystallinity in the films.[33] The transistor is an enhancement-mode device with a threshold voltage of approximately −0.45 V. The transistor has a subthreshold swing of 0.15 V per decade, a maximum transconductance of 13.5 μS, and an on/off current ratio of 4000.

P(VPA-AA), like other electrolytes, focuses the electric field caused by a potential applied between two electrodes (in our case, the gate and the channel) within EDLs very close to each electrode. For small negative gate voltages, protons are attracted to the gate electrode, and form a Helmholtz layer. At the channel, deprotonated P(VPA-AA) anions close to positively doped $P(T_0T_0TT_{16})$ chains form an EDL (sketched in Figure 10.3b). The doping level in the channel reflects the density of excess anions at the $P(T_0T_0TT_{16})$-P(VPA-AA) interface resulting from the depletion of protons. Thus, the conduction through the channel can be controlled by the gate potential.

10.4   PHYSICAL MECHANISMS

10.4.1   ELECTROCHEMICAL DOPING VS. FIELD-EFFECT

An obvious question is whether electrochemistry, that is, doping of the *bulk* of the semiconducting polymer film, causes current modulation in the channel, or whether the channel is opened by an analogous process at the *interface* with the polyanionic P(VPA-AA). The following arguments suggest that the latter is the case. First, the polyanionic chains cannot penetrate the semiconducting polymer film (channel) when the gate is negatively biased because they are effectively immobile (Figure 10.5a). Second, the acidic character of P(VPA-AA) decreases the concentration of hydroxide (OH^-) anions (naturally found due to the dissociation of water) in humid air, which could conceivably penetrate into the semiconducting layer. Third, charge consumption in semiconducting polymer, such as regioregular poly (3-hexylthiophene) (P3HT) electrochemical cells typically occurs over a time-scale of seconds or hundreds of milliseconds (depending on the length-scales and the concentration of ions involved),[36] while the P3HT OFETs gated with P(VPA-AA) discussed here exhibit transient times of the order of 50 μs.[37,38] For similar channel dimensions, P3HT-based electrochemical transistors respond slowly, requiring up to several seconds, to a gate voltage step of −0.6 V.[13] The resulting slow response of the transistor, due to ion migration in the semiconductor, typically implies a variation of the transfer characteristic for different gate voltage sweep rates. Such dependence is not observed in the EDLC-OFET. Fourth, the trends and nature of the current-voltage characteristics of this device agree well with "traditional" OFET characteristics, with the exception that the applied voltage necessary for device operation is markedly lower.

The transient response of the EDLC-OFET, using P3HT as semiconductor, is reported with a pure P(VPA-AA) layer (black line, Figure 10.4). Several observations demonstrate the absence of electrochemistry in the bulk of the P3HT layer: (1) there is no slow increase in current after turning on ($V_G = -1$ V) the transistor; (2) the current after turn off ($V_G = 0$ V) is similar to the off current obtained before turning the transistor on; and (3) after applying a positive bias to the gate ($V_G = +1$ V), the drain

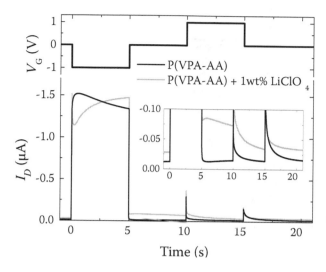

FIGURE 10.4 Chronoamperometric response for a P3HT transistor ($L = 7$ μm, $W = 200$ μm) with a 54-nm thick P(VPA-AA) gate insulator (black line) and a P(VPA-AA) gate insulator "contaminated" with 1wt% of LiClO$_4$ (gray line). The plot shows the response in drain current at constant drain voltage when gate voltage pulses of −1 V and +1 V are applied. The insets show the applied gate voltage and an enlargement of the drain current. All measurements are recorded on samples in humid air (relative humidity ~40%) at room temperature.

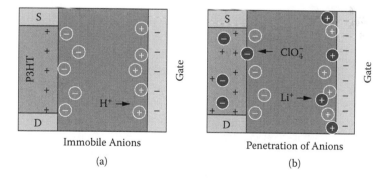

FIGURE 10.5 Sketches of the transistor to illustrate (a) the formation of the electrical double layers with P(VPA-AA) as gate insulator and (b) the penetration of perchlorate anions in P3HT.

current (after the capacitive peak) remains as low. Notice that the capacitive peak current obtained after the +1 V pulse has a slow decay due to ionic current charging the EDLCs at the gate ($V_G = +1$ V) and drain electrode ($V_D = −1$ V). Hence, for a pure P(VPA-AA) layer, the electrochemical doping of P3HT is focused at the interface with the polyanionic P(VPA-AA); in other words, the channel opens due to the formation of an EDL at the conjugated polymer-polyelectrolyte interface (Figure 10.5a).

In order to show the detrimental effect of small anions on the transistor response and study whether the mechanism involved for the EDLC-OFET is bulk electro-chemistry, a transistor with 1wt% LiClO$_4$ blended into the P(VPA-AA) film was studied (gray line, Figure 10.4). The large amount of P(VPA-AA) in the film still leads to a fast rise and fall in the drain current. After the capacitive current peak following the rise, the additional salt leads to a progressive increase in current over several seconds attributed to the enlargement of the conducting channel resulting from the penetration of perchlorate anions into the P3HT layer (Figure 10.5b).[36,39] When the transistor is turned off after having been on for 5 s, the off current is three times larger than in the original off state (before the applied gate pulses), which is likely due to the stability of P3HT$^+$-ClO$_4^-$ in the channel. To completely undope the channel and restore the low conductivity of the neutral semiconducting polymer, a positive potential of +1 V needs to be applied to the gate. This behavior is observed in electrochemical transistors, but is not present in the EDLC-OFET, for which the current level of the off state is independent of the history of the gate bias.

10.4.2 HOLE VERSUS ION MOTIONS

For conventional dielectrics, the polarization is governed by electronic and dipolar relaxations that are tremendously fast. Hence, the limiting mechanism governing the opening of the transistor channel originates from the transport of charge carriers from the injecting contacts. The transit time (τ) of the charge carriers in the channel depends on the channel length (L) according to:[35]

$$\tau \approx \frac{L^2}{\mu V_D}$$

(10.2)

where μ is the charge carrier mobility of the semiconductor and V_D is the source-drain voltage. This explains the traditional picture: shorter channels give faster transistors. However, electrolyte-gated OFETs involve slower polarization mechanisms of the gate insulator. The time required to form the EDL along the channel is limited by either the migration of mobile cations within the polyelectrolyte, or the transport of the holes in the semiconductor,[40] or the charging of the parasitic capacitors that are defined by the overlap between the gate, source, and drain electrodes.[41] The first alternative has been repeatedly suggested for electrolyte-gated OFETs.[14,28] The time constant of the RC circuit constituted by the resistor in series with the parasitic capacitor at the source (C < 1 nF) is so small that the influence of the third phenomenon can be neglected in these measurements.

For the OFETs studied here, switching characteristics are measured for several different OFETs with channel lengths ranging from ~0.2 μm up to 42 μm, and the results are given in Figure 10.6.[37] In that figure, the stray capacitive current decay has been eliminated (see Section 10.5.2), such that the current transient measured actually represents the current evolution passing in the channel. Figure 10.6a shows that the switching rate initially is dependent on the channel length, that is, OFETs with longer channels exhibit slower switching, but that the switching rate finally becomes

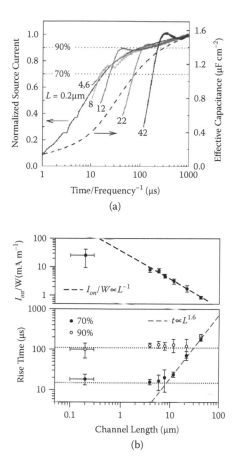

FIGURE 10.6 Switching characteristics for P(VPA-AA) OFETs with different channel lengths. (a) Normalized source current responses for different channel lengths (solid lines) and the effective capacitance as a function of inverted frequency (dashed line). The characteristics shown are averaged curves from measurements on at least three different devices. (b) Source current density reached after 1 ms, that is, on-current density, (upper panel) and rise times for 70% and 90% of the response (lower panel) as a function of channel length. The values were extracted from measurements on numerous OFETs and are averaged for different channel lengths. The error bars indicate standard deviation. All OFETs have a 55-nm thick P(VPA-AA) layer.

independent of the channel length as the normalized current curves coincide. These two regimes are more clearly seen in Figure 10.6b, which shows the extracted rise times as a function of the channel length. The rise time for 70% of the response is nearly proportional to L^2 for large channel lengths ($L > 9$ μm), indicating that the switching current in this regime is limited by the hole transport in the semiconductor channel. For shorter channel lengths ($L < 9$ μm) on the other hand, the rise time is essentially constant, which suggests that the switching current instead is limited by the migration of cations within the electrolyte. Consequently, at the transition point (L = ~9 μm), the holes cross the channel as fast as the electrolyte gets polarized via ionic motions. The rise time including 90% of the response behaves in the same manner, but the transition between the two regimes occurs at a larger channel length (L = ~30 μm).

Figure 10.6a also shows the effective capacitance as a function of the inverted frequency for a metal-electrolyte-semiconductor-metal stack having exactly the same materials and layer thicknesses as the analyzed OFETs. The capacitance curve clearly harmonizes with the current curves for the OFETs with shortest channel lengths, which supports the conclusion that the switching speed of these OFETs is ultimately limited by the migration of ions within the P(VPA-AA) layer. Note that most of the polarization of the electrolyte takes place within just a few tens of microseconds, which allows for fast switching of the OFETs.

10.4.3 Insulator Polarization Mechanisms

Sulfonated polyelectrolytes are known to be hygroscopic; for instance, PSSNa has a water uptake capacity reaching up to 80 wt% at 80% RH.[42] In the presence of water, this polymer swells and the sulfonic acid groups dissociate into fixed (SO_3^-) and mobile (H^+ in the PSSA case) charges. The absorbed water molecules modify the time scale of the various polarization mechanisms of the polyelectrolyte. To study the impact of the ionic conductivity on the polarization of the polyelectrolyte, the level of the RH is varied between 10% and 80% RH. Capacitors made of a thin PSSA layer (80 nm) sandwiched between two titanium electrodes are fabricated and characterized with impedance spectroscopy at various RH levels (10% to 80% RH) at room temperature (20°C). Experimental data (symbols) of the effective capacitance per area and the phase angle are given in Figure 10.7 for different RH levels together with the fit from an impedance model (solid lines).[32] In the periphery of the RH interval studied here, that is, from 10% to 20% RH and from 70% to 80% RH, respectively, the effective capacitance and the phase angle are almost indistinguishable. As the RH increases, the three different frequency regimes (Section 10.2, Figure 10.2), that is, (1) dipolar relaxation, (2) ionic migration, and (3) electric double-layer formation, all shift toward higher frequencies. Note that the capacitive frequency regime at high frequencies (1) is clearly present only for the 10% to 30% RH conditions in the frequency interval analyzed. Upon decreasing the frequency, the transition from dipolar to ionic relaxation is accompanied with a significant increase of the effective capacitance. The transition frequency where $\theta(f) = -45°$ at the low frequency flank, reflecting the formation of the electric double-layers at the two PSSA/titanium interfaces, takes place at relatively low frequencies as the water content of the polyelectrolyte is reduced. At 60% RH, the formation of double-layers occurs at 300 kHz and an electric double-layer behavior is consequently observed over the major part of the frequency interval. In addition, for all RH levels (10% to 80% RH) the effective capacitance reaches a plateau of approximately 20 μF cm^{-2} at 100 Hz.

The 80-nm thick PSSA layer is then used as an electron-insulating layer in OFETs as described in Section 10.3. For channel lengths < 4 μm, the time response of those OFETs is limited by the polarization of the polyelectrolyte.[32] The normalized transient characteristics for an EDLC-OFET with L = 3 μm are given in Figure 10.8a

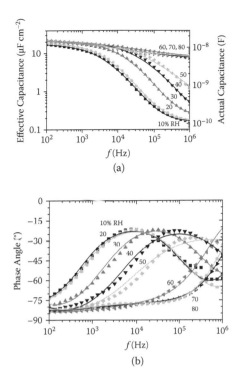

(a)

(b)

FIGURE 10.7 (a) The effective capacitance (per area on the left axis and the actual measured effective capacitance on the right axis) and (b) the phase angle vs. the frequency for different levels of the RH (10% to 80% RH) for a polyelectrolyte EDLC (an 80-nm thick PSSA layer sandwiched between two titanium electrodes). The symbols represent experimental data points, while the solid lines represent the fit of the experimental data to an impedance model.[32]

for different levels of the RH (10% to 50% RH). The expected evolution of $I_{\mathrm{D,Sat}}$ vs. time is given for an EDLC-OFET in Figure 10.8b for different RH levels using the experimental data of the effective capacitance ($C_{\mathrm{Eff}}(f)$, right axis in Figure 10.8b) as the effective capacitance per area of the gate-insulator (C_i) in Equation (10.1). To enable a comparison of the expected current evolution with the actual experiments, given in Figure 10.8a, the constants and parameters of Equation (10.1) are chosen to match the true values of the experiments ($W = 200$ μm, $L = 3$ μm, $\mu = 4 \times 10^{-3}$ cm^2 V^{-1}s^{-1} at $V_G = -1$ V and $V_T = -0.26$ V). Starting with long time-scales, that is, time scale at which the EDLs are formed, the saturated current levels are expected to merge (Figure 10.8b) for all RH levels because the value of the EDLC is not significantly affected by the %RH. This agrees well with the experimental data. The exact saturated current levels of the expected and experimental currents are, however, not identical. This is most likely due to the fact that the P3HT layer present in the OFETs exhibits different surface and charging characteristics as compared to the metal electrode used in the EDLCs and also that the charge carrier mobility value used in the prediction of the drain current might be too high.

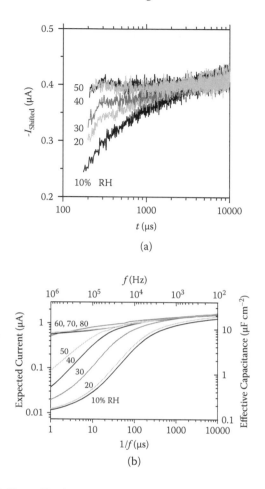

FIGURE 10.8 (a) Normalized current transient illustrating the on switch for an EDLC-OFET at 10% to 50% RH. (b) The expected evolution of the saturation current, calculated from Equation (10.1) using the experimental results of the effective capacitance per area of the polyelectrolyte capacitor (right axis) as C_i, vs. time at 10% to 80% RH.

At intermediate time-scales, the ionic relaxation is dominant and varies significantly with humidity. The currents are expected to rapidly decrease at different rates depending on the %RH and to level out to a saturated value for the 10% to 50% RH levels (Figure 10.8b). This is also in agreement with the transient measurement of the OFETs. Note that at higher RH levels (60% to 80% RH), electric double-layers are formed on a very short time-scale (MHz). Note the significant electrolysis of water observed at humidity levels above 50% RH.[38] At the short time-scales, the currents are expected to be limited by dipolar relaxation, which is characterized by the significantly lower value of the effective capacitance in Figure 10.8b.

Thus, for EDLC-OFETs that are limited by the polarization of the polyelectrolyte (short channels), the three frequency regions that exist for the EDLCs alone can also be used to predict and describe the transient characteristics for EDLC-OFETs. The fast formation of the electric double-layers observed in humid conditions (60% to 80% RH) with a capacitance of 20 μF cm^{-2} at 1 MHz demonstrates the potential of using an electrolyte as gate-insulator to obtain EDLC-OFETs that operate at 1 MHz and 1 V.

10.5 POTENTIAL FOR PRINTED ELECTRONICS

10.5.1 COMPATIBILITY WITH PRINTING TECHNIQUES

According to the electrolyte model presented here, a polymer transistor gated via a polyanionic electrolyte should essentially have the same operating voltages independently of the thickness of the electrolyte layer, and of the position of the gate electrode. In the EDLC-OFET illustrated in Figure 10.9a, the gate electrode of a neighboring transistor is utilized to modulate the conductance of the channel (gate lines separated by 1.1 mm). The drain current in the output characteristics (Figure 10.9b) is lower than that observed from the vertically gated EDLC-OFET,[26] possibly because the induced charges in the P3HT are located over the source electrode rather than in the channel. Adjusting the position of the gate electrode relative to the channel may improve this. In Figure 10.9c, a much thicker P(VPA-AA) layer (thickness = ~200 μm) is cast from a drop onto the P3HT film. The output characteristics display clear current modulation (Figure 10.9d) for the same operating voltage as the OFETs with a P(VPA-AA) thickness of 80 nm. The drain current is much lower than in the two other transistor configurations. This could be explained by a weaker capacitive coupling between the gate and channel, due to poorer contact between the metal wire and the electrolyte and the decreased area of the gate electrode. These two experiments demonstrate that the device behavior is robust with respect to the alignment and patterning of the electrolyte and gate electrode, a typical challenge when taking printed organic electronics to production. Dewetting is expected to be an issue when printing either the organic semiconductor solution of weak polarity on the hydrophilic polyelectrolyte film or the polar polyelectrolyte solution on the hydrophobic organic semiconductor film. This problem can be solved by using an adhesive layer of an amphiphilic copolymer made of a segment of conjugated polymer and a segment of the polyelectrolyte. The use of ionic liquids or gels based on ionic liquids as electrolytes is attractive due to their more hydrophobic character, thus eliminating the wetting problems when printing. [29]

10.5.2 PRINTING AND MINIATURIZATION

Many of potential applications for OFETs like, for example, in radio frequency identification (RFID) tags and addressing backplanes in displays, put requirements on the switching speed and the current throughput of the OFETs. Ideally, the switching time and the drain current (I_D) are proportional to L^2/μ and μ/L, respectively, where L is the channel length and μ is the charge carrier mobility.[35,40] Major efforts have been devoted to improve the charge-carrier field-effect mobility in the organic polymers and a mobility of 0.1 to 0.6 cm^2 V^{-1} s^{-1} has been reached in several material systems.[43,44] In addition, different routes to achieve short OFET channels, typically less

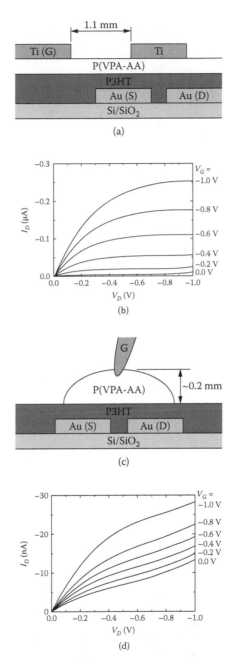

FIGURE 10.9 Alternative gating of EDLC-OFETs. (a) Schematic cross-section of a laterally gated transistor manufactured as the EDLC-OFET. (b) Typical output curves obtained for the device shown in (a) ($L = 9$ μm, $W = 200$ μm). (c) Schematic cross-section of a transistor with a hemispherical P(VPA-AA) gate insulator formed on top of the P3HT layer by letting a drop of polyelectrolyte solution dry. A metal wire was pushed against the droplet to act as the gate electrode. (d) Output curves obtained for the device sketched in (c) ($L = 43$ μm, $W = \sim500$ μm).

than 1 μm, have extensively been explored during the past decade, using techniques such as photolithography,[45] electron-beam lithography,[46-48] nanoimprint lithography,[49] underetching,[50] embossing,[51] microcontact printing,[52] and inkjet printing.[41,53,54] However, many of the previously reported OFETs with submicron channels have displayed deteriorated current-voltage characteristics that deviate from the ideal, long-channel behavior, and are referred to as short-channel effects. Typically for an OFET that exhibits short-channel effects, the threshold voltage decreases and the subthreshold current becomes dependent on the drain voltage (V_D), which reduces the on/off current ratio.[49,55] Moreover, the output curves are often superlinear and do not saturate above the so-called pinch-off voltage. Both these effects reduce the usefulness of the transistor, particularly in circuits.

The increased longitudinal electric field that is induced along the submicron channel, in between the source and drain contacts, is generally considered to be the main contribution to the short-channel effects.[35,55] When this component becomes comparable to the transverse electric field, induced by the gate potential, the resulting field in the channel becomes a two-dimensional function of the channel position. Hence, the standard transistor equations are not valid because they assume that the electric field in the channel is a one-dimensional function of the channel position (the so-called gradual channel approximation). For very high longitudinal fields, a space-charge limited current (SCLC) that flows through the bulk of the semiconductor layer dominates the drain current,[49,55] which results in diode-like output characteristics. High contact resistance has a similar effect on the output characteristics, but generally dominates the drain current only at the lowest drain voltages.[45,49] Another proposed mechanism for the short-channel effects is that the mobility depends on the drain voltage.[56]

The short-channel effects can be suppressed by making the transverse electric field much larger than the longitudinal electric field. One route to increase the transverse electric field is simply to reduce the film thickness, for example, by using self-assembled monolayers (SAMs).[46] The longitudinal electric field can be reduced by inserting an insulating mesa-like structure between the source and drain electrodes.[47] Also, lower off-currents and improved on/off current ratios can be achieved by reducing the thickness of the semiconductor layer.[48]

Due to the ability to form high electric fields in the Helmholtz double layers, polyelectrolytes, such as P(VPA-AA), can be used as gate insulators to suppress short-channel effects in submicron channel OFETs based on regioregular poly(3-hexylthiophene) (P3HT). The OFETs have a top-gate bottom-contact configuration (Figure 10.10a). The source and drain electrodes were fabricated by self-aligned inkjet printing (SAP) of gold nanoparticles[41,54] giving typical channel lengths of 200 nm.

In order to show the features of the polyelectrolyte compared to traditional dielectrics, SAP OFETs were fabricated with non-electrolytic poly(methyl methacrylate) (PMMA) gate insulators, and their current-voltage characteristics are given in Figure 10.10b,c. For the 200-nm thick PMMA gate dielectric ($C = 17$ nF cm^{-2} at 1 kHz), pronounced short-channel behavior is observed with the absence of saturation (Figure 10.10b). Reducing the PMMA layer thickness down to ~55 nm ($C = 55$ nF cm^{-2} at 1 kHz) increases the transversal electric field (up to ~10^{-8} V m^{-1}

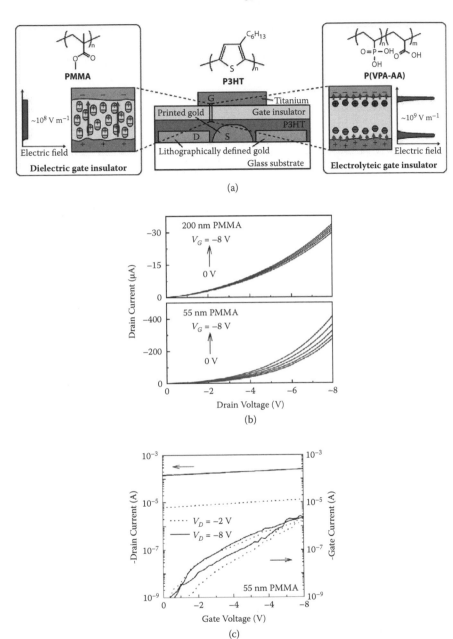

FIGURE 10.10   Electrical characteristics of submicron channel length SAP OFETs having dielectric (left) and electrolytic (right) gate insulators, PMMA and P(VPA-AA), respectively. (a) Schematic cross-sections and illustrations of the electric field distribution inside the insulator when a negative potential is applied to the gate electrode for SAP OFETs with dielectric PMMA and electrolytic P(VPA-AA) gate insulators. The molecular structures of P3HT, PMMA, and P(VPA-AA) are also shown. (b) Output and (c) transfer characteristics for SAP OFETs with 55-nm and 200-nm thick PMMA gate insulators.   **(Continued)**

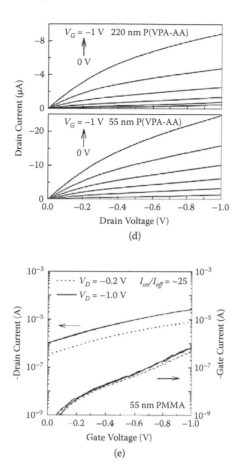

FIGURE 10.10 (Continued) (d) Output and (e) transfer characteristics for SAP OFETs with 55-nm and 220-nm thick P(VPA-AA) gate insulators. The channel lengths and channel widths are ~200 nm and 500 μm, respectively, for all the OFETs.

at 8 V). However, such a thin PMMA layer does not aid the recovery of long-channel behavior. The fabrication of even thinner gate insulator layers, without significant gate leakage currents, is either extremely difficult or very time consuming. The superlinear shape of the output curves also leads to poor drain current modulation as seen in the transfer characteristics (Figure 10.10c).

The electrical characteristics of typical electrolyte-gated SAP OFETs are given in Figure 10.10d,e. The initial linear dependence of the drain current on the drain voltage in the output characteristics (Figure 10.10d), at very small V_D values, suggests low source and drain contact resistances. For slightly higher drain voltages, the output curves bend off as the pinch-off point is approached, in agreement with long-channel OFETs. The short-channel effects are greatly reduced as compared to the PMMA-gated SAP OFETs (Figure 10.10b), which suggests that the transverse electric field in the channel is substantially higher in the polyelectrolyte-gated OFET (estimated to $\sim 10^9$ V m^{-1} at 1 V) despite the lower gate voltage (V_G). Interestingly, a similar level of saturation is observed for different thicknesses (55 and 220 nm) of the polyelectrolyte layer. The magnitude of both the capacitance of the electrolyte and the electric field at the semiconductor surface is essentially independent of the thickness of the electrolyte layer. This specific property of polyelectrolytes is a clear advantage compared to traditional gate insulators, that is, dielectrics, when it comes to fabrication of the insulator layers. Polyelectrolytes allow for using thick layers that can be manufactured via printing techniques.

Notice that above the pinch-off, I_D does not saturate completely and increases linearly with V_D. This is indicative for channel-length modulation (CLM), which usually is the first short-channel effect that is observed when the channel length is reduced.[55] This effect becomes apparent in transistors that are operated above pinch-off when the width of the depletion region at the drain contact (ΔL), which increases with V_D, gets comparable to L. The effective channel length then reduces to $L - \Delta L$, and consequently, since I_D is inversely proportional to the channel length, I_D increases with V_D. CLM can be accounted for to a first-order approximation by adding the term $(1 + \lambda V_D)$, where λ is the CLM parameter, to the saturation current. CLM has been observed in polyelectrolyte-gated OFETs with much longer channel lengths ($L \leq 42$ μm).

Figure 10.10e shows the gate voltage dependence of the drain and gate currents (I_G) at two different drain voltages. The on/off current ratio is ~ 25, and the extracted threshold voltage is close to 0 V, which is approximately 0.3 V higher than what is normally found for several microns long channels. The threshold voltage shift could indicate another short-channel effect called threshold voltage roll-off, which shifts the transistor operation from enhancement mode to depletion mode as the channel length is reduced.[55] The low threshold voltage, together with the CLM, also explains the rather large difference between the drain currents obtained for the two different drain voltages. Moreover, the gate current is virtually independent of V_D, and is generally much lower than I_D.

As mentioned earlier, the switching speed of the transistor is expected to increase when the channel length is shortened. The transient response of an electrolyte-gated SAP OFET to a square voltage pulse of 1 ms applied at the gate is given in Figure 10.11. The source current (I_S) is measured for two different drain voltages, 0 V and −1 V, where the former response only contains the current arising from the charging of the gate capacitor (the exponentially decaying current in Figure 10.11a). The effective current through the channel is extracted by simply taking the difference between the two recorded currents (Figure 10.11b,c). The rise times, including 70% and 90% of the response are 10 μs and 42 μs, respectively; and the corresponding fall times 3.3 μs and 11 μs, respectively. These rise and fall times are significantly shorter than what has been previously reported for electrolyte-gated OFETs,[14,28] thus expanding the field of applications for these kinds of OFETs.

10.6   ELECTRONIC CIRCUITS

Many of the targeted applications for printed organic electronics require powering from printed batteries,[6] solar cells,[7] or via electromagnetic induction.[8] Power sources

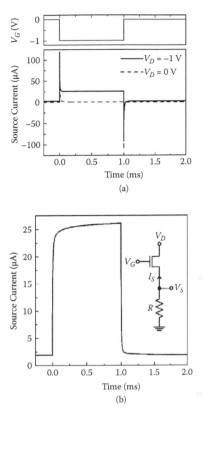

(a)

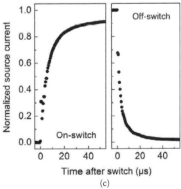

(b)

(c)

FIGURE 10.11 Switching characteristics of a SAP OFET having a 55-nm thick P(VPA-AA) gate insulator. (a) Applied gate voltage and recorded source currents through the resistor (R) when applying 0 V and −1 V to the drain electrode. (b) Source current response obtained after subtracting the charging current and circuit schematic of the measurement setup. (c) The rise and fall of the source current shown in more detail. The current is normalized to 1 when on and to 0 when off.

possible to integrate with printed electronics can typically supply an entire electronic system with a voltage ranging from 1 V to only a few volts at a current often limited to less than 1 mA. For proper operation and also for rapid updating of the printed electronics, the clock frequency should be at least 100 Hz.[8] Hence, tremendous efforts are presently devoted to bring down simultaneously the driving voltage of printable organic transistors and the delay time.

Polyelectrolytes are tested as a gate insulator material in OFETs and transistor circuits. The OFETs have a bottom-contact, top-gate configuration (Figure 10.12a). The source and drain electrodes, only 3.5 µm wide, are interdigitated in order to minimize the vertical overlap with the gate electrode, hence reducing the parasitic stray capacitance (Figure 10.12b). The polythiophene derivative, $P(T_0T_0TT_{16})$, is used as a semiconductor,[30] and the polyanionic electrolyte P(VPA-AA) is chosen as the gate insulator. The cut-off frequency can be estimated to be about 8 kHz, while assuming that the gate capacitor is defined by the vertical overlap of the channel, source, drain, and gate electrodes, giving a gate capacitance of 0.26 nF.

Figure 10.13 shows the electrical characteristics, circuit schematics and micrographs of two integrated inverters using different load transistor designs: one with the gate connected to the drain (saturated load), and another with the gate connected to the source (depleted load). The channel widths of the driver and load transistors are 15 mm and 3 mm, respectively, in the saturated load inverter and vice versa in the depleted load inverter. The channel length is 2.5 µm for all transistors. Both inverters possess good transfer characteristics and voltage amplification, that is, $|\partial V_{OUT}/\partial V_{IN}| > 1$, which is a prerequisite for using the inverters in logic circuits. The depleted load inverter has the best static performance with a nearly full voltage swing and a high voltage gain. However, the load transistor is always "off" because the threshold voltage is negative. The resulting relatively high channel resistance of the load transistor, in relation to the capacitances, will limit the switching speed of the low-to-high transition. The design is more useful for a load transistor that operates in depletion-mode.

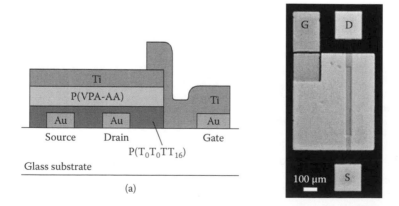

FIGURE 10.12 (a) Schematic cross-section of the OFET and the vertical interconnect. (b) Micrograph of a TFT with indicated gate (G), drain (D), and source (S) contacts.

In the saturated load inverter, the load transistor instead operates in the saturation regime, that is, it is always "on," and hence does provide a relatively higher current throughput.

Seven-stage ring oscillators with output buffers are fabricated using the saturated load inverter design and a micrograph image of the circuit is displayed in Figure 10.14. The ring oscillator shows stable oscillations for supply voltages down to −0.9 V. The output signal for the ring oscillator driven with a supply voltage of −1.5 V possesses an oscillation frequency of 235 Hz (Figure 10.14b). The voltage swing is large and corresponds well to the voltage measured for individual inverters. The evolutions of the oscillation frequency and the calculated stage delay time vs. the supply voltage are reported in Figure 10.14c. The stage delay time decreases with increasing supply voltage, and a minimum stage delay of 300 µs is obtained at −1.5 V supply voltage. The signal delay is within a factor three from the estimated cut-off frequency. To our knowledge, these values are the lowest signal delay times reported for organic circuits at supply voltages of 1.5 V or less.[57]

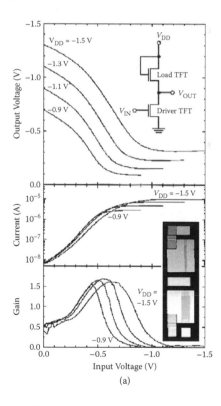

(a)

FIGURE 10.13 Characteristics of inverters with electrolyte-gated TFTs. Output voltage, current, and signal gain as a function of input voltage for supply voltages between −0.9 V and −1.5 V for (a) a saturated load inverter with a load-to-driver channel width ratio of 5 The insets show circuit schematics and micrographs of the two different inverters. The channel length is 2.5 µm for all transistors. **(Continued)**

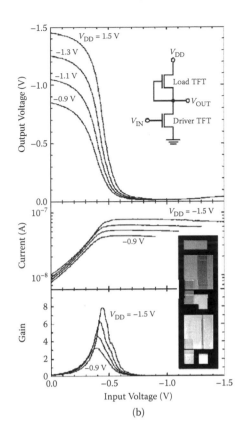

(b)

FIGURE 10.13 (Continued) Characteristics of inverters with electrolyte-gated TFTs. Output voltage, current, and signal gain as a function of input voltage for supply voltages between –0.9 V and –1.5 V for (b) a depleted load inverter with a driver-to-load channel width ratio of 5. The insets show circuit schematics and micrographs of the two different inverters. The channel length is 2.5 μm for all transistors.

10.7 SYNOPSIS

We have shown that a polyanionic proton conductor can be used as a high capacitance layer in an OFET, significantly decreasing the operating voltage (< 1V). The extraordinarily high capacitance originates from EDLs that are formed at the electrolyte-(semi)conductor interfaces or, more precisely, from the very short distance between the ionic and electronic charges. The short distance also enables a high transversal electric field at the electrolyte-semiconductor interface, which has been shown to suppress short-channel effects in printed OFETs. Employing a polyelectrolyte as a gate insulator also eases design and manufacturing requirements. Controlling the alignment of the gate electrode and the thickness of the gate insulator becomes less critical since the nanometer-scale "capacitor" spontaneously forms at the semiconductor-insulator interface as the gate is biased. This enables us to make low-voltage and relatively fast transistors in robust device structures, which is of great importance for flexible electronics.

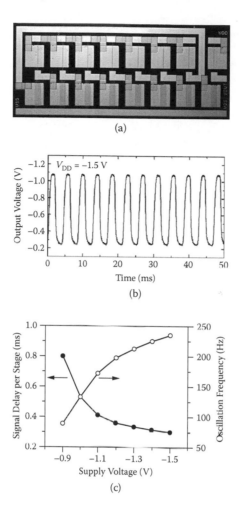

(a)

(b)

(c)

FIGURE 10.14 Characteristics of a seven-stage ring oscillator with electrolyte-gated TFTs. The inverters are of saturated load type ($W_{(driver\ OFET)} = 15$ mm; $W_{(load\ OFET)} = 3$ mm). The channel length is 2.5 µm for all TFTs. (a) Micrograph of a ring oscillator circuit. (b) Output voltage signal for a supply voltage of -1.5 V, showing an amplitude of 0.8 V and a period of 4.2 ms. (c) Oscillation frequency and calculated signal propagation delay as a function of supply voltage.

Compared to electrochemical transistors and other electrolyte-gated OFETs, polyelectrolyte-gated OFETs show a fast polarization of ions in the electrolyte, thus leading to switch on and off in approximately 10 µs. Electrochemical doping of the polymer semiconductor bulk is prevented because only immobile anions are present at the semiconductor-polyelectrolyte interface. Relatively fast switching speeds have also been demonstrated for OTFTs gated with ionic gels. The switching speed only is limited by the polarization of the polyelectrolyte in OFETs with short channel lengths ($L < 5$ to 9 µm). Three different polarization mechanisms are identified: (1) at high frequencies (> 120 kHz at 10% RH) a region attributed to dipolar relaxation of the polyelectrolyte, (2) at intermediate frequencies (800 Hz to 120 kHz at 10% RH) a region attributed to ionic relaxation, and (3) at low frequencies (< 800 Hz at 10% RH) a region attributed to the EDL formation at the polyelectrolyte/metal electrode interfaces. The fast formation of the EDLs observed at humid conditions (60% to 80% RH) with a capacitance of 10 μFcm^{-2} (at 1 MHz) demonstrates the potential of electrolyte-gated OFETs for operations at 1 MHz and 1 V.

The benchmark ring oscillators fabricated from polyelectrolyte-gated OFETs display a minimum stage delay of 300 μs at –1.5 V supply voltage. Hence, polyelectrolyte-gated OFETs are promising candidates for use in low-voltage printed electronic applications such as RFID labels, smart printings, and possibly distributed diagnostics. The high capacitance and quick polarization of the polyanionic electrolyte allows the fabrication of digital transistor circuits operating at a clock frequency of 1 kHz and at 1 V drive voltage, thus enabling powering from printed batteries, organic solar cells, and electromagnetic induction. Employing a semiconductor material with even higher charge carrier mobility should further improve the speed of the transistor circuits, perhaps enabling a clock frequency around 10 kHz.

10.8   ACKNOWLEDGMENTS

The authors gratefully acknowledge the Swedish Foundation for Strategic Research (OPEN), the Knut and Alice Wallenberg Foundation, Brains and Bricks, VINNOVA, the Royal Swedish Academy of Sciences, and the Swedish Research Council for financial support.

REFERENCES

1. H. Sirringhaus, T. Kawase, R. H. Friend, T. Shimoda, M. Inbasekaran, W. Wu, E. P. Woo, Science 2000, 290, 2123.
2. L.-L. Chua, J. Zaumseil, J.-F. Chang, E. C. W. Ou, P. K. H. Ho, H. Sirringhaus, R. H. Friend, Nature 2005, 434, 194.
3. J. H. Burroughes, D. D. C. Bradley, A. R. Brown, R. N. Marks, K. Mackay, R. H. Friend, P. L. Burns, A. B. Holmes, Nature 1990, 347, 539.
4. M. Berggren, T. Kugler, Physics World 2001, 14, 20.
5. T. Kawase, H. Sirringhaus, R. H. Friend, T. Shimoda, Adv. Mater. 2001, 13, 1601.
6. J. Z. Wang, Z. H. Zheng, H. W. Li, W. T. S. Huck, H. Sirringhaus, Nature Mater. 2004, 3, 171.
7. L. A. Majewski, R. Schroeder, M. Grell, Adv. Funct. Mater. 2005, 15, 1017.
8. A. Facchetti, M. H. Yoon, T. J. Marks, Adv. Mater. 2005, 17, 1705.
9. M. Halik, H. Klauk, U. Zschieschang, G. Schmid, C. Dehm, M. Schutz, S. Maisch, F. Effenberger, M. Brunnbauer, F. Stellacci, Nature 2004, 431, 963.
10. L.-L. Chua, P. K. H. Ho, H. Sirringhaus, R. H. Friend, Adv. Mater. 2004, 16, 1609.
11. P. Bergveld, Sens. Actuators B 2003, 88, 1.
12. S. Chao, M. S. Wrighton, J. Am. Chem. Soc. 1987, 109, 2197.
13. S. Chao, M. S. Wrighton, J. Am. Chem. Soc. 1987, 109, 6627.
14. M. Taniguchi, T. Kawai, Appl. Phys. Lett. 2004, 85, 3298.
15. M. J. Panzer, C. R. Newman, C. D. Frisbie, Appl. Phys. Lett. 2005, 86, 103503.
16. M. J. Panzer, C. D. Frisbie, J. Am. Chem. Soc. 2005, 127, 6960.
17. M. J. Panzer, C. D. Frisbie, Adv. Funct. Mater. 2006, 16, 1051.

18. M. J. Panzer, C. D. Frisbie, Appl. Phys. Lett. 2006, 88, 203504.
19. A. S. Dhoot, J. D. Yuen, M. Heeney, I. McCulloch, D. Moses, A. J. Heeger, Proc. Natl. Acad. Sci. U.S.A. 2006, 103, 11834.
20. D. Nilsson, M. Chen, T. Kugler, T. Remonen, M. Armgarth, M. Berggren, Adv. Mater. 2002, 14, 51.
21. N. Robinson, P.-O. Svensson, D. Nilsson, M. Berggren, J. Electrochem. Soc. 2006, 153, H39.
22. T. G. Bäcklund, H. G. O. Sandberg, R. Österbacka, H. Stubb, Appl. Phys. Lett. 2004, 85, 3887.
23. H. G. O. Sandberg, T. G. Bäcklund, R. Österbacka, H. Stubb, Adv. Mater. 2004, 16, 1112.
24. M. J. Panzer, C. R. Newman, C. D. Frisbie, Appl. Phys. Lett. 2005, 86, 103503.
25. E. Said, X. Crispin, L. Herlogsson, S. Elhag, N. D. Robinson, and M. Berggren, Appl. Phys. Lett. 2006, 89, 143507.
26. L. Herlogsson, X. Crispin, N. D. Robinson, M. Sandberg, O.-J. Hagel, G. Gustafsson, M. Berggren, Adv. Mater. 2007, 19, 97.
27. S. Ono, S. Seki, R. Hirahara, Y. Tominari, J. Takeya, Appl. Phys. Lett. 2008, 92, 103313.
28. J. Lee, M. J. Panzer, Y. He, T. P. Lodge, C. D. Frisbie, J. Am. Chem. Soc. 2007, 129, 4532.
29. J. H. Cho, J. Lee, Y. Xia, B. Kim, Y. He, M. J. Renn, T. P. Lodge, C. D. Frisbie, Nature Mater. 2008, 7, 900.
30. S. Mitra, A. K. Shukla, S. Sampath, J. Power Sources 2001, 101, 213.
31. E. Bard, L. Faulkner, Electrochemical Methods: Fundamentals and Applications, Wiley, New York, 2001.
32. O. Larsson, E. Said, M. Berggren, X. Crispin, Adv. Funct. Mater., 2009, 19, 3334.
33. H. Ohkita, S. Cook, Y. Astuti, W. Duffy, S. Tierney, W. Zhang, M. Heeney, I. McCulloch, J. Nelson, D. D. C. Bradley, J. R. Durrant, J. Am. Chem. Soc. 2008, 130, 3030.
34. S. Scheinert, G. Paasch, Phys. Status Solidi A 2004, 201, 1263.
35. S. M. Sze, Physics of Semiconductor Devices, Wiley, New York, 1981.
36. T. Johansson, N. K. Persson, O. Inganas, J. Electrochem. Soc. 2004, 151, 119.
37. L. Herlogsson, Y.-Y. Noh, N. Zhao, X. Crispin, H. Sirringhaus, M. Berggren, Adv. Mater. 2008, 20, 4708.
38. E. Said, O. Larsson, M. Berggren, X. Crispin, Adv. Funct. Mater. 2008, 18, 1.
39. I. N. Hulea, H. B. Brom, A. J. Houtepen, D. Vanmaekelbergh, J. J. Kelly, E. A. Meulenkamp, Phys. Rev. Lett. 2004, 93, 166601.
40. L. Bürgi, R. H. Friend, H. Sirringhaus, Appl. Phys. Lett. 2003, 82, 1482.
41. Y.-Y. Noh, N. Zhao, M. Caironi, H. Sirringhaus, Nature Nanotech. 2007, 2, 784.
42. F. Toribio, J. P. Bellat, P. H. Nguyen, M. Dupont, J. Colloid Interface Sci. 2004, 280, 315.
43. H. Sirringhaus, P. J. Brown, R. H. Friend, M. M. Nielsen, K. Bechgaard, B. M. W. Langeveld-Voss, A. J. H. Spiering, R. A. J. Janssen, E. W. Meijer, P. Herwig, D. M. de Leeuw, Nature 1999, 401, 685.
44. I. McCulloch, M. Heeney, C. Bailey, K. Genevicius, I. MacDonald, M. Shkunov, D. Sparrowe, S. Tierney, R. Wagner, W. Zhang, M. L. Chabinyc, R. J. Kline, M. D. McGehee, M. F. Toney, Nature Mater. 2006, 5, 328.
45. C.-A. Di, G. Yu, Y. Liu, X. Xu, Y. Song, Y. Wang, Y. Sun, D. Zhu, H. Liu, X. Liu, D. Wu, Appl. Phys. Lett. 2006, 88, 121907.
46. J. Collet, O. Tharaud, A. Chapoton, D. Vuillaume, Appl. Phys. Lett. 2000, 76, 1941.
47. J. Z. Wang, Z. H. Zheng, H. Sirringhaus, Appl. Phys. Lett. 2006, 89, 083513.
48. K. Tsukagoshi, F. Fujimori, T. Minari, T. Miyadera, T. Hamano, Y. Aoyagi, Appl. Phys. Lett. 2007, 91, 113508.
49. M. D. Austin, S. Y. Chou, Appl. Phys. Lett. 2002, 81, 4431.
50. S. Scheinert, T. Doll, A. Scherer, G. Paasch, I. Horselmann, Appl. Phys. Lett. 2004, 84, 4427.
51. N. Stutzmann, R. H. Friend, H. Sirringhaus, Science 2003, 299, 1881.

52. M. Leufgen, A. Lebib, T. Muck, U. Bass, V. Wagner, T. Borzenko, G. Schmidt, J. Geurts, L. W. Molenkamp, Appl. Phys. Lett. 2004, 84, 1582.
53. C. W. Sele, T. Von Werne, R. H. Friend, H. Sirringhaus, Adv. Mater. 2005, 17, 997.
54. N. Zhao, M. Chiesa, H. Sirringhaus, Y. Li, Y. Wu, B. Ong, J. Appl. Phys. 2007, 101, 064513.
55. J. N. Haddock, X. Zhang, S. Zheng, Q. Zhang, S. R. Marder, B. Kippelen, Org. Electron. 2006, 7, 45.
56. L. Wang, D. Fine, D. Basu, A. Dodabalapur, J. Appl. Phys. 2007, 101, 054515.
57. L. Herlogsson, M. Cölle, S. Tierney, X. Crispin, M. Berggren, Adv. Mater, 2010, 22, 72.

Index

W